Indian Alchemy

SOUTH ASIA RESEARCH

Current Series Editor
Donald R. Davis

Previous Series Editor
Martha Selby
A Publication Series of
The University of Texas South Asia Institute
and
Oxford University Press

THE EARLY UPANISADS
Annotated Text and Translation
Patrick Olivelle

INDIAN EPIGRAPHY
A Guide to the Study of Inscriptions in Sanskrit, Prakrit, and the other Indo-Aryan Languages
Richard Salomon

A DICTIONARY OF OLD MARATHI
S. G. Tulpule and Anne Feldhaus

DONORS, DEVOTEES, AND DAUGHTERS OF GOD
Temple Women in Medieval Tamilnadu
Leslie C. Orr

JIMUTAVAHANA'S *DAYABHAGA*
The Hindu Law of Inheritance in Bengal
Edited and Translated with an Introduction and Notes by Ludo Rocher

A PORTRAIT OF THE HINDUS
Balthazar Solvyns & the European Image of India 1740-1824
Robert L. Hardgrave

MANU'S CODE OF LAW
A Critical Edition and Translation of the Manava-Dharmasastra
Patrick Olivelle

NECTAR GAZE AND POISON BREATH
An Analysis and Translation of the Rajasthani Oral Narrative of Devnarayan
Aditya Malik

BETWEEN THE EMPIRES
Society in India 300 BCE to 400 CE
Patrick Olivelle

MANAGING MONKS
Administrators and Administrative Roles in Indian Buddhist Monasticism
Jonathan A. Silk

SIVA IN TROUBLE
Festivals and Rituals at the Pasupatinatha Temple of Deopatan
Axel Michaels

A PRIEST'S GUIDE FOR THE GREAT FESTIVAL
Aghorasiva's Mahotsavavidhi
Richard H. Davis

DHARMA
Its Early History in Law, Religion, and Narrative
Alf Hiltebeitel

POETRY OF KINGS
The Classical Hindi Literature of Mughal India
Allison Busch

THE RISE OF A FOLK GOD
Viṭṭhal of Pandharpur
Ramchandra Chintaman Dhere
Translated by Anne Feldhaus

WOMEN IN EARLY INDIAN BUDDHISM
Comparative Textual Studies
Edited by Alice Collett

THE RIGVEDA
The Earliest Religious Poetry of India
Edited and translated by Stephanie W. Jamison and Joel P. Brereton

CITY OF MIRRORS
The Songs of Lālan Sāī
Translated by Carol Salomon
Edited by Saymon Zakaria and Keith E. Cantú

IN THE SHADE OF THE GOLDEN PALACE
Alaol and Middle Bengali Poetics in Arakan
Thibaut d'Hubert

TO SAVOR THE MEANING
The Theology of Literary Emotions in Medieval Kashmir
James Reich

THE OCEAN OF INQUIRY
Niścaldās and the Premodern Origins of Modern Hinduism
Michael S. Allen

A LASTING VISION
Dandin's Mirror *in the World of Asian Letters*
Edited by Yigal Bronner

EAST OF DELHI
Multilingual Literary Culture and World Literature
Francesca Orsini

MINOR MAJESTIES
The Paḻuvēṭṭaraiyars and their South Indian Kingdom of Paḻuvūr, 9th-11th centuries A.D.
Valérie Gillet

INDIAN ALCHEMY
Sources and Contexts
Edited by Dagmar Wujastyk

Indian Alchemy

Sources and Contexts

Edited by
DAGMAR WUJASTYK

OXFORD
UNIVERSITY PRESS

OXFORD
UNIVERSITY PRESS

Oxford University Press is a department of the University of Oxford.
It furthers the University's objective of excellence in research, scholarship,
and education by publishing worldwide. Oxford is a registered trade mark of
Oxford University Press in the UK and in certain other countries.

Published in the United States of America by Oxford University Press
198 Madison Avenue, New York, NY 10016, United States of America.

CIP data is on file at the Library of Congress

ISBN 9780197805800

DOI: 10.1093/9780197805831.001.0001

Printed by Marquis Book Printing, Canada

The manufacturer's authorized representative in the EU for product safety is
Oxford University Press España S.A., Parque Empresarial San Fernando de Henares,
Avenida de Castilla, 2 – 28830 Madrid (www.oup.es/en or product.safety@oup.com).
OUP España S.A. also acts as importer into Spain of products made by the manufacturer.

MIX
Paper | Supporting responsible forestry
FSC® C103567
FSC
www.fsc.org

Contents

List of Illustrations

Acknowledgments

Research for this volume was undertaken for the AyurYog Project (www.ayur yog.org) and made possible through funding from the European Union's Horizon 2020 research and innovation program under grant agreement no.639363.

Contributor Bios

Keith Edward Cantú is Lecturer on South Asian Religious Traditions at Harvard Divinity School and Postdoctoral Fellow at the Center for the Study of World Religions. His research especially focuses on South Asian yoga and tantra, and the interface between Sanskrit and Indic vernacular languages, especially Bengali and Tamil, and on the connected histories of yoga and music in global esoteric currents. He is the author of *Like a Tree Universally Spread: Sri Sabhapati Swami and Śivarājayoga* and co-editor of *City of Mirrors: Songs of Lālan Sā̃i*, both published with Oxford University Press.

Priyanka Chorge is an ayurvedic doctor with a Bachelor of Ayurvedic Medicine and Surgery (B.A.M.S., 2008) and an M.D. Ayurveda from the Sumathi Bhai Shah Ayurved Mahavidyalay in Pune, India. She is the Founder and Managing Director of Ayuryogalaya in Hamburg. Dr Chorge is currently also pursuing an MA in South Asian studies at the University of Hamburg.

James Mallinson is an Indologist whose work focuses on the history of yoga and its practitioners. He is the Boden Professor of Sanskrit at the University of Oxford.

Patricia Sauthoff is a historian of medieval Sanskrit religious and alchemical traditions. She teaches in the Department of History at Hong Kong Baptist University and is the author of *Illness and Immortality: Mantra, Maṇḍala, and Meditation in the Netra Tantra* (2022).

David Gordon White is Distinguished Emeritus Professor of Religious Studies at the University of California–Santa Barbara and Associate Research Fellow at the Centre d'Etudes Sud-Asiatiques et Himalayennes (Paris). Among his publications are *The Alchemical Body: Siddha Traditions in Medieval India* (1996), *Kiss of the Yogini: "Tantric Sex" in Its South Asian Contexts* (2003), *Sinister Yogis* (2009), *The Yoga Sutra of Patanjali: A Biography* (2014), and *Dæmons Are Forever: Contacts and Exchanges in the Eurasian Pandemonium* (2021).

Dagmar Wujastyk is an Indologist whose work focuses on the history of South Asian alchemy and medicine. She is Associate Professor in South Asian Studies in the Department of History, Classics, and Religion at the University of Alberta. She is the author of *Well-Mannered Medicine* (Oxford University Press, 2012) and editor of several volumes on Indian medical and alchemical themes.

Dominik Wujastyk's primary languages of research are Sanskrit and Pali. He has published research on Indigenous Sanskrit grammar, Indian codicology, the history of science and medicine in India, and the history of Yoga. He is Professor Emeritus and former Singhmar Chair at the Department of History, Classics, and Religion at the University of Alberta, Canada.

Introduction

The Indian Alchemical Tradition

Dagmar Wujastyk

In the tenth century CE, alchemists in India began to compose works in Sanskrit about their craft, defining both its goals and the methods with which to achieve them. This new genre of literature set practices that were previously the stuff of legend within a framework of an authoritative discipline with formal boundaries. The practitioners of this discipline considered goals such as making gold, lengthening lifespan, attaining superhuman powers, and entering transcendent states achievable through technical means and especially through a curated use of enhanced substances. Indian alchemists developed a set of procedures they believed would transform inorganic and organic materials into substances with the power to transmute base metals to noble ones, and ordinary humans into perfected beings. At the same time, alchemical products could also be applied in this-worldly ways, such as in the treatment of disease, or to counter age-related conditions, forming a subdiscipline of iatrochemistry that over time became the dominant feature of Indian alchemy.

In the European and Middle Eastern traditions, the term "alchemy" is used for a range of ideas and practices, many of which center on the transmutation of materials and especially on the making of gold. The metallurgical aspect of these traditions is reflected in the very name of alchemy, which probably derives from the Greek *chemeia* or *chumeia*—the "art of melting (metals)," via the later Arabic *al-kīmiyā'*.[1] The Indian traditions used differing terms for their discipline. The earliest of the Indian alchemical works, the tenth-century *Rasahṛdayatantra* ("Heart of Mercury"), called it "the knowledge

[1] For a historical overview of European and Middle Eastern alchemical traditions, see Principe 2013. For a discussion of the etymology of the term *chemeia/chumeia*, see chapter 1, and especially page 24.

Dagmar Wujastyk, *Introduction* In: *Indian Alchemy*. Edited by: Dagmar Wujastyk, Oxford University Press.
 DOI: 10.1093/9780197805831.003.0001

of mercury" (*rasavidyā*), and also referred to the discipline as a *śāstra*, an established body of knowledge. The use of the term *rasavidyā* is quite common in Sanskrit alchemical literature, as is the denomination "the discipline of mercury" (*rasaśāstra*).[2] The eleventh- or twelfth-century *Rasārṇava* ("Ocean of Mercury") also referred to "the doctrine of mercury" (*rasavāda*), contrasting it with other doctrines.[3] In his eleventh-century Arabic work on the customs of India, Al-Biruni described alchemical practice with the term *rasāyana*, a term long established in the Indian medical tradition.[4] Al-Biruni was mistaken in picking this term as the title of the whole discipline. The Sanskrit alchemical works themselves did not use this term to describe the discipline as a whole. Their use of the term *rasāyana* refers to the final phase of the alchemical process: the period in which the mercurial elixir is taken and the regimen that accompanies its intake.[5] As may be obvious from the terms used for the discipline, the alchemical texts all share one aspect in particular, namely a focus on mercury: this alchemy is the discipline of mercury.

Alchemy before *Rasaśāstra* Literature: From *Dhātu-* to *Rasavāda*

The new genre of literature did not mark the beginning of alchemy on the Indian subcontinent: A number of sources tell us of the presence of alchemical practices centuries earlier. Waley (1932) listed references to alchemical practices in Chinese Buddhist texts, some of which might date to around the fourth or fifth century. All of these make reference to the making of gold, while mercury is not mentioned in any of them.[6] In the seventh century, the

[2] Keyword searches in the *Digital Corpus of Sanskrit* reveal that reference to *rasavidyā* is, for example, found in the *Rasārṇava*, *Rasaratnākara*, *Rasendracintāmaṇi*, *Rasaratnasamuccaya*, and *Rasamañjarī*, while variants of "the discipline of mercury," i.e., *rasaśāstra*, *rasasya śāstra*, or *rasendrasaya śāstra*, are used in the *Rasahṛdayatantra*, *Rasārṇava*, *Rasaratnākara*, *Rasendracūḍāmaṇi*, *Rasaratnasamuccaya*, *Rasamañjarī*, and *Rasaprakāśasudhakara*. In the following, I will use the term *rasaśāstra* as an umbrella term for Indian alchemy.

[3] *Rasārṇava* 1.44. See White in this volume on these doctrines.

[4] Ayurvedic literature traditionally divides the discipline of medicine into eight subject areas. One of these subject areas is *rasāyana*. In medical literature, the term describes herbal rejuvenation, anti-ageing, and life-prolonging treatments, i.e., vitalization therapy. See Maas 2017 on the difficulties of translating this term. In the case of the alchemical tradition, the *rasa* in *rasāyana* is typically understood to denote mercury, though it may also refer to the essential element of any substance.

[5] The alchemical elixir regimen parallels the structure of the much older *rasāyana* regimen of the ayurvedic tradition. See Wujastyk 2017 on the details of these parallels.

[6] Waley's lists of texts include a passage from the *Avataṃsakasūtra*, four passages from the *Mahāprajñāpāramitopadeśa*, a passage from the *Mahāyānasaṃgrahabhāṣya*, and one passage from the *Abhidharmamahāvibhāṣā* (Waley 1932: 1102–1103). It is not entirely clear whether the passages in question go back to Indian original versions, or whether they were later additions, as no Sanskrit

Chinese Buddhist pilgrim Xuanzang told stories of Nagarjuna in his *Record of the Western World*, in which Nagarjuna is described as "well practised in the art of compounding medicines; by taking a preparation (pill or cake) he nourished the years of his life for many hundreds of years so that neither the mind nor appearance decayed."[7] In a different story, Nagarjuna produces gold "by moistening all the great stones with a divine and superior decoction (medicine or mixture)."[8] These abilities are not connected with each other in Xuanzang's retelling, however, and therefore should not be understood as describing two aspects of one alchemical discipline. Rather, they form part of Nagarjuna's attributed repertoire of special abilities as a bodhisattva.

There is further textual evidence for alchemy in Buddhist and more specifically Vajrayāna milieus, which also gives witness to connections between alchemy and yogic practices. The *Kālacakratantra* ("Teaching on the Wheel of Time," ca. eleventh century) and its commentary, the *Vimalaprabhā* ("Stainless Light," eleventh century), describe yogic practices that consist of meditations, visualizations, recitations of mantras, breathing exercises, and postures. The second chapter of the *Kālacakratantra* juxtaposes these yogic practices with a series of alchemical recipes that involve the use of mercury and other metals and minerals, with many more similar formulae in its fifth chapter.[9] Some of the passages of the *Vimalaprabhā* are similar to ones in the *Rasārṇava*.[10]

During a period in which Vajrayāna thought and practices were absorbed into the yoga practices of Śaiva Nāth practitioners, alchemical metaphors started to appear in works on yoga to describe the experience of yogis. Perhaps the earliest of these works is the *Amṛtasiddhi* ("The Attainment of Immortality"), an eleventh-century Sanskrit work on yoga that was composed in a Vajrayāna milieu.[11] The work uses alchemical metaphors to

originals of any of these texts survive. The noted passages may therefore represent Chinese ideas about alchemy rather than Indic ones.

[7] Beal 1884: 210 and Rongxi 1996: 273. It should be noted that this does not necessarily connect Nagarjuna with the alchemical tradition, but could instead refer to his knowledge of ayurvedic vitalization therapy, whose tonics were also meant to prolong life and retain youthfulness.

[8] Beal 1884: 216 and Rongxi 1996: 276.

[9] See Fenner 1979:146–149, and 159–184 for a translation of the relevant recipes of chapter 2 and chapter 5 in the Tibetan version of the *Kālacakratrantra* and the corresponding commentary of the *Vimalaprabhā*. The Tibetan version seems to have a different verse count than the Sanskrit version. The relevant verses in Fenner's translation in chapter 2 begin with verse 125, which corresponds to verse 131 in the Sanskrit version in Upadhyaya 1986: vol. 1, 239.

[10] See White 1996: 384–385n124 on some similar passages in the *Rasārṇava* and the *Vimalaprabhā*.

[11] Mallinson, 2019: 3 and Mallinson 2020: 416–420. Birch (2024) argues for a similar date for the *Amaraugha* (a Sanskrit text on yoga attributed to Gorakṣanātha), which contains the same kinds of alchemical metaphors as the *Amṛtasiddhi*. Birch suggests that the *Amaraugha* and *Amṛtasiddhi* draw on a common source rather than that the *Amaraugha* borrows from the *Amṛtasiddhi* (pp.14–15).

denote bodily processes. For example, the term *māraṇa*, which in alchemical treatises denotes the "killing" or calcination of materials, is used to describe the stilling or stopping of the breath. Some of the *Amṛtasiddhi*'s metaphors were carried over into later works. However, later authors of Sanskrit texts on yoga often failed to understand alchemical metaphors that originated in the *Amṛtasiddhi*.[12]

We also find particularly early mention of alchemy (Prakrit: *dhāuvvāo*) in Jain narratives. A very brief mention in Saṅghadāsa's *Vasudevahiṇḍī* ("Vasudeva's Wanderings") from before the sixth century describes the making of gold from iron pyrites and *rasa*, which may or may not be mercury, undertaken by an ascetic.[13] Haribhadra's Prakrit version of the *Dhūrtākhyāna* ("Stories of Rogues," eighth century CE) describes a well full of mercury that has the power to transform one thousand times its weight of base metals into gold.[14]

There is also a more extensive description of alchemical practice in the eighth-century Jain novel *Kuvalayamālā* ("The Garland of Blue Water-Lilies") by Uddyotana Sūri. In its chapter "The Return to Ayodhya," there is a section in which the hero, prince Kuvalayacandra, meets a group of alchemists in a forest in the Vindhya Mountains. Observing the failure of the alchemists to produce alchemical gold and their resulting dejection, he steps in and helps them. The description of the alchemists' activities in this narrative reflects a certain degree of metallurgical knowledge: the alchemists are credited with a knowledge of what happens when you heat and melt metals or expose them to flames and how metals react to each other and to other reagents. Here, we also see the use of some technical terms that reappear in later alchemical literature, though mercury is conspicuously missing from the named ingredients.[15] Later, Prince Kuvalayacandra lectures Princess Kuvalayamālā on the practice of alchemy, notably presenting alchemy as one of the disciplines (*kalā*) that an educated person should know. He notes that only persons of superior moral and cognitive qualities can practice the discipline successfully and proposes a typology of three types of alchemists: "The man who is familiar with the procedure of mixing (the substances) together is

[12] Mallinson and Szántó 2022 and Mallinson in this volume.

[13] The passage is discussed in Balbir 1990: 136–137 and 143–144. Also see Balbir 1992 for other references to alchemy in Jain literature.

[14] See Balbir 1990: 135.

[15] The terminology is discussed in Balbir 1990. A translation of the entire passage is available in Chojnacki and Nagarajaiah 2018: vol. 2, 875–885.

(the man known as) one 'conversant with alchemical operations' (*kiriyāvāī*); he who has skill in fixing (all substances right up to) mercury is (the man who is called) a 'master alchemist' (*ṇariṃdo*). The man who knows how to take the metal out of a soil (propitious to this end) and blows it with the aid of alkalis is called, obviously, by all an 'alchemist' (*dhāuvvāī*), so it is said."[16]

As in the earlier Buddhist accounts, alchemy is presented as a practice for transmuting substances into gold. In the eleventh-century *Kathākośaprakaraṇa*, Jineśvara differentiates between a doctrine of metals and minerals (*dhāuvāo*) and a doctrine of mercury (*rasavāo*).[17] Both disciplines seem aimed at producing precious metals, and gold in particular. As Balbir (1990: 134) noted, there is no mention of the transmutation of the human being in the early Jain narratives: the craft is associated with attaining wealth. The *Kalyāṇakāraka* ("The Cause of Welfare"), a ninth-century medical work composed by the Digambara monk Ugrāditya, presents a departure from the focus on gold. In its twenty-fourth chapter, it describes the making of mercurial tonics for the attainment of health and longevity. The chapter features technical terms that appear in later alchemical texts and also describes a number of apparatuses and their use in alchemical procedures. Later Jain alchemical works from the thirteenth century on seem to carry this new focus on medicine forward, while also maintaining an interest in the older metallurgical aspects of alchemy.[18]

Outside of Jain literature, the seventh-century Sanskrit novel *Kādambarī* by Bāṇabhaṭṭa should be mentioned. It refers to a doctrine of metals (*dhātuvāda*) in its satirical description of a tantric practitioner. The same passage describes the tantric's failed use of a *rasāyana*. White (1996: 49) translated *rasāyana* with "mercurial elixir." However, the term *rasāyana* here may refer to a medicinal tonic rather than to a mercurial elixir, given its intended use as a vermifuge.[19] In any case, Bāṇabhaṭṭa's account brings the themes of Śaiva tantric mantra practice, magical acts (*ṣatkarman*), digging

[16] *jo kuṇaī joya-juttiṃ kiriyāvāī tu so bhave puriso* | *jo uṇa baṃdhai ṇiuṇo rasaṃ pi so bhaṇṇai ṇariṃdo* || *jo geṇhiūia dhāuṃ khettāo dhamai khār-juttīe* | *so kira bhaṇṇai payaḍaṃ dhāuvvāī juṇe sayale* || (Upadhye 1959: 197). Translation of chapter 4.3.195.29–30 in Chojnacki and Nagarajaiah 2018: vol. 2, 884. The term translated as "blows" (*dhamai*) should rather be "smelts." The Prakrit *kiriyāvāī* is the Sanskrit *kriyāvādin*, and *dhāuvvāī* corresponds to the Sanskrit *dhātuvādin*.

[17] See Balbir 1990: 143.

[18] See Sikdar 1980: 10.

[19] Several of the early ayurvedic works (*Suśrutasaṃhitā*, *Aṣṭāṅgahṛdayasaṃhitā*, and *Aṣṭāṅgasaṃgraha*) mention the vermifugal action of some of their *rasāyana* tonics, while the earliest alchemical works do not seem to associate their elixirs with anthelmintic action.

for treasure, and *rasāyana* into context with one another, anticipating their integration within later Śaiva alchemical works.[20]

Finally, Al Biruni's eleventh-century account of alchemy in India makes note of longevity practices alongside aurifaction, though Al Biruni seems to confuse the medical and alchemical *rasāyana* and also confesses to not actually having witnessed alchemical practice himself.[21]

None of these early sources (bar perhaps the *Kalyāṇakāraka* and the *Kālacakratantra* and *Vimalaprabhā*), however, were written by alchemists. They are narrative accounts of outsiders to the discipline and are more often than not critical of the craft and its practitioners. The new texts, by contrast, gave voice to how alchemists themselves wished their discipline to be perceived. Their literary representation of alchemy created an orthodoxy of alchemical practice, but also staked a claim for alchemy as an authoritative body of knowledge.

Mercurial Alchemy (*Rasavāda, Rasaśāstra*)

The new alchemical literature that emerged around the beginning of the second millennium CE is characterized by a shift in perspective from outsider to insider, but also by a shift in aims from gold-making to spiritual attainments. In these works, mercury was set at the center of the discipline, with subsidiary substances aiding its processing into an alchemical elixir. The introduction of mercury as the key to transmutation is mysterious, because there are no significant mercurial deposits on the Indian subcontinent.[22] And yet, the new works gave no hint of being indebted to foreign ideas or supplies. Gold-making was still very much within the purview of Indian alchemists, but was portrayed as a litmus test for the mercurial elixir, proving its transmutative potency, rather than as an end in itself.

The early *rasaśāstra* works displayed some differences in how they portrayed the mercurial discipline, and the later works echoed and then developed these perspectives, adding their own distinctive characteristics

[20] On this story and other mentions of alchemy in Indian narrative literatures, see Baldissera 2015: 21–22. See Cantú in this volume for the idea of digging for treasure in the *Rasaratnākara*.

[21] Sachau 1910: 187–189, and 188 specifically on Al-Biruni not having witnessed alchemical practice himself.

[22] See White 1996: 62. White speculated that the introduction of mercury is due to influence and supply from China. He noted, however, that there might have been traces of mercury in volcanic areas of India that might have sufficed for the practice of Indian alchemy. See White 1996: 377n63.

and emphases. One of the significant differences between the early alchemical works lies in their narrative frameworks. The *Rasahṛdayatantra*'s author Govinda dedicated his summary of the craft in 506 verses to the king of the Kirāṭas, whom he referred to as a master on the subject of mercury (*rasācārya*). Here, one expert is speaking to another, soberly delivering a learned précis on a practical discipline with soteriological aims that is familiar to both. The stated soteriological aims include liberation in a living body (*jīvanmukti*) and the attainment of the Absolute or Supreme Spirit, here named *brahman*. Notably, the *Rasahṛdayatantra* does not display strong sectarian affiliation: it is the only early work within this body of literature that is not emphatically Śaiva.[23] The *Rasahṛdayatantra* is not a guide to practice, but rather a statement on what the discipline's activities consist of and what their purpose is. It stakes a claim for alchemy as an authoritative body of knowledge, a *śāstra*, grounded in expert knowledge and established methods.

The *Rasārṇava*, on the other hand, is firmly placed in a Śaiva, or more specifically Kaula, setting. It is presented as a dialogue between Śiva and Pārvati, in which Śiva instructs Pārvati on alchemy as esoteric knowledge. Here, mercury is a manifestation of Śiva as the Absolute in the form of his semen, and must be worshiped.[24] The *Rasārṇava* is a much longer treatment of the subject (2,300 verses) and includes many elements entirely missing in the *Rasahṛdayatantra*, such as the auspicious positioning and set-up of the laboratory, lists of instruments used in alchemical operations, the initiation of the alchemist, the desired characteristics of the persons involved in the alchemical endeavor, and the use of mandalas, mantras, and propitiations of deities to ensure the success of all alchemical operations. Many of these elements signal a Śaiva tantric worldview.[25] The *Rasārṇava* makes explicit mention of the discipline's affiliation with the Kaula tradition, and White (1995: 282) noted that the *Rasārṇava* must be read as "an attempt to situate the way of mercury or the mercurial science (*rasavidyā*) within the broader contexts of tantric and Hindu doctrine and practice."

[23] For example, it describes mercury as "produced by Hara," i.e., Śiva, but also states that it "conquers like Hari," i.e., Viṣṇu (RHT 1.2), while also making reference to the attainment of *brahmabhāva*—absorption into Brahman/the Absolute (RHT 1.26).

[24] On the subject of mercury as the semen of Śiva, see the translation of *Rasārṇava*, chapter 1 in White 1995: 281 and in this volume.

[25] See White 1996: 142–145 and in this volume on some of the tantric elements in the *Rasārṇava* and other alchemical works.

Several later works carry the narrative framework of a dialogue among various forms of the god and goddess forward, portraying alchemy as an esoteric science only accessible to initiated adepts. These works typically feature tantric elements as well.[26] White (1996: 145–146) noted a division between alchemical works that focus on special powers (*siddhi*) and bodily immortality as the alchemist's ultimate goal and are closely linked to Hindu Tantras and those that emphasize increased longevity as a means to the higher end of liberation (*mukti*), following Śaiva Siddhānta metaphysical and soteriological orientations and adopting elements of ayurvedic vitalization therapy (*rasāyana*).[27] Broadly, this distinction reflects the differences between the *Rasahṛdayatantra* and the *Rasārṇava*, suggesting two lineages of alchemical traditions starting with these early works. However, the overall picture is not entirely clear cut. Later works often borrowed from both early treatises, quoting or paraphrasing them, elaborating on their themes, or echoing their structure. They also referred to texts that predominantly display the characteristics of the other group. This means that we cannot discern clear lineages on the basis of the texts. Rather, we can note that alchemical writers seem to have had access to and made use of the different early streams of alchemical writing and synthesized them. This intertextuality is one aspect of the literature that lets us identify *rasaśāstra* as one single, albeit not entirely unified, alchemical tradition. More importantly, beyond ideological affiliation, all *rasaśāstra* works are deeply connected through a shared vision of alchemical activities.

The Alchemical Worldview

There are three basic principles that are found in all Sanskrit alchemical works. The first is the belief in the possibility of transmutation, namely that one substance can become another, and a human being can be transformed

[26] Works that employ the format of a conversation between god and goddess include the *Rasārṇavakalpa* (Bhairava and Pārvatī), the *Rasopaniṣat* (Śiva and Pārvatī), the *Ānandakanda* (Bhairava and Bhairavī), the *Gorakṣasaṃhitā*'s *Bhūtiprakaraṇa* (Śrīkaṇṭha and Śakti), the *Rasamañjarī* (Śiva and Pārvatī), and the *Kākacaṇḍeśvarīmata* (Śiva and Kākacaṇḍī).

[27] White (1996: 145) includes the *Rasārṇava*, the *Kākacandeśvarīmata*, the *Rasendramaṅgala*, the *Rasaratnākara*, and the *Gorakṣasaṃhitā*'s *Bhūtiprakaraṇa* in the first group, and suggests they implicitly follow the doctrines of texts such as the *Kaulajnānanirṇaya* of Matsyendra, the *Kubjikāmata*, the *Kulārṇavatantra*, and the *Rudrayāmala* and *Brahmayāmala*. Works that fall into his second category include the *Rasahṛdayatantra*, *Rasopaniṣat*, *Rasendracūdāmaṇi*, *Rasaprakāśasudhākara*, *Rasaratnasamuccaya*, and *Ānandakanda*.

into an entity of an entirely different ontological status. The second principle shared by all the *rasaśāstra* works is that mercury is key to effecting transmutation. This differentiates *rasaśāstra* or *rasavāda* from the earlier *dhātuvāda*. And the third is that mercury and its subsidiary substances must be processed in particular ways in order to make an effective transmutative elixir.

Most of the materials employed in alchemical processes are already found in ayurvedic literature, where they are ingredients in medicinal compounds. This includes the metallic and non-metallic substances that constitute the main ingredients of alchemical elixirs. However, the ayurvedic works that predate alchemical literature do not emphasize the use of metallic minerals. Mercury in particular barely plays a role for ayurvedic therapy before the eleventh century.[28]

However, the main innovation in alchemical works lies not so much in which materials are used, but in the concept of the need to process materials before they can be used. Materials must be detoxified, their essence extracted, and their potency harnessed. Here, we find a divide in how substances are used: one group constitutes the main ingredients, typically metallic and non-metallic minerals. These materials are subsidiary to mercury and, just like mercury, must be purified and potentiated or have their essences extracted before they are added to the mercurial elixir. The processed substances have a dual function in the making of the elixir: on the one hand, they enhance mercury's potency and efficacy; on the other hand, they affect the elixir's overall action.

The other group of substances consists of auxiliary materials that aid in the processing of the main materials, including mercury, to make them fit for use (i.e., to detoxify them, or to extract their essence). These auxiliary materials are often herbal and typically do not themselves need to undergo processing before their application. However, some materials are used as both subsidiary and auxiliary substances, so that this boundary is not fixed.

The alchemical innovation is that plants and minerals become a kind of medicine for other materials, a concept not found in early ayurvedic works. This principle is the basis of a prominent feature of all *rasaśāstra* works, namely the progressive sets of alchemical operations. Those dedicated to refining mercury were called *saṃskāra* (refinements) or *rasakarman* (mercurial operations). There is no specific overarching term for the procedures

[28] On this topic, see Wujastyk 2019 and 2013.

dedicated to the refinement of subsidiary substances such as mica and sulfur, though the relevant procedures may be broadly grouped into cleansing (*śodhana*), essence extraction (*sattvapātana*), and calcination (*māraṇa*). The descriptions of these procedures and the terminology used for them aid in establishing both an understanding of the grouping of materials and a chronological order among alchemical works.[29]

Descriptions of these procedures constitute the bulk of all *rasaśāstra* works. Thus, judging purely by the proportion of text dedicated to this aspect of the discipline, we must conclude that Indian alchemy was first and foremost a technical and practical craft: a discipline of doing and making. A typical day in the life of an alchemist would likely have been spent dealing with materials, acquiring or gathering them, grinding, mixing, heating, and distilling or sublimating them. Alchemists needed access to and technical know-how about alchemical instruments and apparatuses. The work would have necessitated tactile and other sensory knowledge, for example to ascertain when a material was ground finely enough, what texture and color a mixture or distillate should have, or how materials interacted and reacted to heat. One should not underestimate the physical aspect of this work, elbow grease clearly being a key factor in making an elixir.[30] A further element of the practice of alchemy lies in the passing of time. If one were to perform all the procedures, making an elixir would probably take months, if not years. An alchemist would thus have had to be able to dedicate substantial amounts of time and effort—and probably also financial means—to the craft.[31] An alchemist's work may have been very similar to that of an artisanal metallurgist, the main difference lying in the final products and their anticipated applications. The early alchemical works ascribed a range of effects to their elixirs, spanning the spectrum from mundane (better health, improved sexual stamina, rejuvenation, prolongevity) to sublime (experiencing

[29] See Hellwig 2014 on the subject of material groups in the Sanskrit alchemical works and how they differ from ayurvedic categorizations. See also Hellwig 2009a on how particular patterns of which procedures are mentioned in alchemical works provide information on the work's chronological positioning within Indian alchemical literature. See also my chapter in this volume on the *saṃskāras* in the *Rasaprakāśasudhākara* and Wujastyk 2019.

[30] However, see the chapter by D. White in this volume, which describes the various actors engaged in alchemical work. The relevant passage from the *Rasārṇava* suggests that there may have been a divide in the alchemical team between those holding theoretical knowledge and those performing the physical work.

[31] None of the alchemical works explicitly address the issue of the finances needed for the practice of alchemy. However, the descriptions of the laboratory, the instruments, and the needed materials imply some wealth on part of the alchemist.

enlightenment, liberation in a living body). This makes alchemy a practical discipline with both worldly and soteriological aims.

But what about alchemy as a religious or spiritual practice? Was the process of making the elixir spiritually significant and part of religious practice? Here, there is no clear-cut answer. The earliest of the alchemical works, the *Rasahṛdayatantra* and *Rasārṇava*, present somewhat different perspectives. The *Rasahṛdayatantra* acknowledges mercury as a special substance with outstanding characteristics that makes it ideal for a life-prolonging elixir (see Chapter 1 in this volume on the culmination of alchemical practice). In its first chapter, the treatise describes three uses of mercury: "Having thickened, it cures disease; bound, it gives liberation; well-killed, it makes immortal. What promotes compassion more than mercury?" *(Rasahṛdayatantra 1.3).*[32] However, in the following verses, the mercurial elixir is not represented as an agent for liberation, but as a means for achieving health and longevity. It is the prolonged, healthy life that will allow practitioners to properly dedicate themselves to spiritual practice and ultimately gain liberation. In this context, mercury is simply a useful substance, which must first be processed to be applied with success. In its nineteenth chapter, the *Rasahṛdayatantra* ascribes to mercury a more direct effect on liberation. However, the treatise does not describe any ritual activity centering on mercury and also makes no mention of devotional acts. The significance of the alchemical operations is simply to make an effective, non-toxic elixir that will have liberative effects.

The *Rasārṇava* echoes the idea of the need to preserve the body through applying mercury. This will then enable a person to first acquire higher knowledge that will in turn allow them to achieve liberation. The treatise then explains that mercury is identical with Śiva: "Mercury (*sūta*) is identical to me, o Goddess. It arises from my every limb. It is my bodily fluid (*deharasa*), which is why it is called *rasa*" (*Rasārṇava* 1.36).[33] This implies that any kind of interaction with mercury constitutes an encounter with the divine. Mercury is the object of worship and also an offering in worship of Śiva: "By viewing, touching, eating, indeed by simply recalling, worshiping, and offering mercury, its six-fold reward is made manifest. The merit [gained] from viewing all the liṅgams in the world—Kedārnāth and

[32] Variations of this verse are found in many of the alchemical works, including the *Rasārṇava*, as well as in the *Haṭhayogapradīpika*. See White 1996: 274–275 and 495n55.

[33] Translation by David Gordon White. See Chapter 2 in this volume. In the first chapter of the *Rasahṛdayatantra*, mercury is twice referred to as an emission of Hara and Gaurī (*haragaurīsṛṣṭi* (1.7 and 1.33). However, these seem to be the only instances in the treatise in which we find this reference. This may be an interpolation.

so on—as well as all others that may exist, is [realized] through the simple viewing of mercury."

"Mercury that has become 'swooned' (*mūrcchita*)[34] when compounded with sandal, camphor, and saffron [is employed] in the worship of Śiva, for the *siddhi* of [eternal] nearness to Śiva" (*Rasārṇava* 1.37–39).[35]

Note, however, that the processing of mercury through the *saṃskāra* is not mentioned as one of the meritorious acts here, nor in the following verses. Chapter 2 of the *Rasārṇava* outlines the setup of the laboratory, and this involves ritual placement of mantras, the tracing of mandalas, and the propitiation of deities to consecrate the ritual space of the laboratory and to start the alchemical work appropriately.[36] Chapter 3 continues the description of preparatory rituals. However, reading carefully, one can see that, while all of this sets alchemical work within a specific ritual context, these ritual actions are separate from the activities of alchemical work proper (i.e., the alchemical operations). The tone of the work changes from chapter 4 onward, turning from ritual technologies to the practical aspects of making the elixirs. The text gives lists of materials and equipment, and then details the alchemical operations.[37] The operations (*rasakarma*) described in the following chapters (especially chapter 10), are not accompanied by ritual activities or acts of worship[38] and do not seem to be understood as ritual activities themselves. In other words, the execution of the various procedures is not presented as a religious activity with spiritual consequences for the practitioner. It is only the intake of the elixir that leads to the spiritual transformation of the person who consumes it (which may be someone different from the person making the elixir). Remove chapters 1–3 and the narrative framework of the dialogue between god and goddess, and you are more or less left with activities that are not ascribed with any specific spiritual significance. There are, however, some exceptions. For example, the practitioner, called an adept (*sādhaka*) in the *Rasārṇava*, is promised the removal of all sins (*pāpa*) and liberation (*mokṣa, mukti*) through performing the "digestion" or

[34] For the technical term *mūrcchita* (also *mūrchana, mūrchā*, etc), also see Chapter 8 in this volume on the *saṃskāra* in the *Rasaprakāśasudhākara*. The associated procedure has the effect of thickening the mercury, but also weakening it.

[35] Translation by David Gordon White. See Chapter 2 in this volume.

[36] See, for example, *Rasārṇava* 1.103 (translation by David Gordon White in Chapter 2 in this volume) on completing these installments and beginning the alchemical work proper.

[37] See Meulenbeld 2000: 675–679 for a useful synopsis of the eighteen chapters of the *Rasārṇava*.

[38] There are a few exceptions, several of which are found in chapter 12, such as the description of how to make "enlivening water" (*saṃjīvanījala*) (*Rasārṇava* 12.232–247). This recipe, which does not involve mercury, requires the incantation of mantras.

amalgamation (*jāraṇa*) stage of the *saṃskāra*.[39] Here, performing alchemy becomes a spiritual activity.

Manual or Pamphlet: The Purpose of Alchemical Literature

It is uncertain whether alchemists actually undertook all the work described in the treatises. This brings up the question of how the alchemical treatises were used by alchemists. The texts examined in this volume all seem to prescribe actions: the predominant verbal modes found in them are the optative and gerundive. However, when one tries to follow the recipes or formulations using the text alone, one quickly finds that crucial information is missing.[40] At the same time, the texts do not seem deliberately opaque, but rather read like ayurvedic works, in that they seem to assume that everyone generally knows how things work, so that one need not express every last detail.[41] Perhaps the texts were not actually meant to be used as a manual, or at least, not on their own. It is possible that they were used as didactic devices by teachers, who provided the missing information. However, one could also interpret the texts as articulating a normative version of the craft. This would imply that the literature was in some sense directed outward, to an audience of other alchemists, but perhaps also non-alchemists, who were to be persuaded of the legitimacy of the discipline. As mentioned, it is also the case that some of the texts present the craft as an esoteric, secret discipline, open only to a select group of initiates. However, the very fact that the methods of the discipline were put in writing in texts composed in the universal scholarly language of Sanskrit suggests otherwise. Some of the texts seem to have been distributed fairly widely and are referenced in medical works.

There is a further element found in some alchemical manuscripts that points to some scholarly or artisanal exchange, rather than a jealous guarding of esoteric secrets: there are diagrams of apparatuses on some manuscripts of the *Rasendramaṅgala* and the *Rasaratnākara*. Similar line drawings of apparatuses are also featured in at least two Perso-Indian manuscripts of

[39] See *Rasārṇava* chapter 11, 2–6.

[40] See the experiments of the AyurYog project on reconstructing alchemical operations according to the texts at http://ayuryog.org/content/alchemy-reconstruction.

[41] In this, the Sanskrit alchemical works are markedly different from the Tamil medico-alchemical works, which employ coded language. On this subject, see Kędzia 2017: 125.

medico-alchemical works.[42] One of these texts mentions the *Rasaratnākara* as well as other Sanskrit alchemical texts, so that we can infer at least some exchange among practitioners of iatrochemistry from different communities or at least a transmission of Sanskrit/Śaiva alchemy to Muslim scientific communities in India. The use of diagrams in the Sanskrit alchemical works is very unusual and may be a distinct feature of the above-mentioned works, rather than of alchemical literature more broadly. This remains an active area of research. The integration of diagrammatic visual materials into these works may be due to influence from Islamicate sciences, though whether the images on the manuscripts of the *Rasendramaṅgala* and the *Rasaratnākara* pre- or postdate those of the Perso-Indian manuscripts is as yet unknown.[43]

From Elixir Alchemy to Iatrochemistry

While the earliest alchemical works already show a knowledge of medicine and specifically of the methodology of medical vitalization therapy (*rasāyana*) in their prescriptions for the intake of the mercurial elixirs, they do not focus on the therapeutic potential of their products outside of the elixir regimen, where the restorative and invigorating tonic effects of elixirs are noted.[44] However, over time, Indian alchemists showed increasing interest in the medical uses of the materials used in alchemical processes and the therapeutic applications of their elixirs. Going beyond the scope of the initial vitalization and aphrodisiac effects of elixirs, alchemists began to extol the ability of their products to treat and cure specific diseases, using categories familiar to the ayurvedic tradition.[45] The perhaps first detailed discussions of disease categories appear in the twelfth- to thirteenth-century *Rasendracūḍāmaṇi* ("The Crest Jewel of Mercury"), whose chapter 6 to 8 focus on medicinal plants, and chapters 9 to 16 discuss the medicinal use

[42] See Speziale 2006 and 2019 on the Perso-Indian manuscripts and Wujastyk 2023 on diagrams in manuscripts of the *Rasendramaṅgala*.

[43] At present, this is a matter of some speculation, as there is presently no study dedicated to the history of diagrams in either Islamicate or Indian alchemy. Stapleton and Azo 1905 discuss a thirteenth-century manuscript of an eleventh-century Persian alchemical work with simple diagrams of apparatuses. Many of the manuscripts of the *Rasendramaṅgala* and *Rasaratnākara* that contain diagrams are undated, but the earliest dated ones go back to the seventeenth century. The topic of diagrams on Indian manuscripts and on alchemical manuscripts more specifically deserves to be the subject of a larger study.

[44] See Dagmar Wujastyk, Chapter 1 in this volume on this subject.

[45] See Wujastyk 2017: 26–27 for elixirs for diseases in alchemical works, and 30 for diseases mentioned in the context of vitalization therapy in ayurvedic works.

of substances such as minerals, gems, metals, and mercury.[46] This marks the beginning of a stronger focus on medicine based on ayurvedic principles in Sanskrit alchemical works. The thirteenth- to fifteenth-century *Rasaratnākara* ("The Jewel Mine of Mercury") dedicates one of its five books (its *Rasendrakhaṇḍa*, "The Section on the Lord of Essences") to the subject of medicine. More than half of the sixteenth-century *Rasaratnasamuccaya* ("Jewel Collection of Mercury") is about the medicinal applications of mercurials and other mineral and herbal medicines, while the sixteenth-century *Rasendrasārasaṃgraha* ("Collection of the Essence of the Lord of Essences") is entirely devoted to medicine. Generally, all later works devote a substantial portion to iatrochemistry, though they may also feature other elements such as the making of artificial gold or the attainment of special powers. For example, the *Rasaprakāśasudhākara* ("Nectar Mine Light on Mercury"), a work from the sixteenth century, is mostly focused on medicine, but also describes all the standard procedures of making a mercury elixir, and gives recipes for making gold, silver, pearls, and coral.

The seventeenth to twentieth centuries saw a decrease in the production of Sanskrit alchemical works. This gap was perhaps filled by alchemical works written in Persian and in vernaculars and by the incorporation of alchemical methods of drug production in ayurvedic works.[47] There was, however, a renaissance of Sanskrit alchemical literature in the late nineteenth and early twentieth centuries, with both new works on *rasaśāstra* being written and print editions of old works being produced. The new works were typically focused on iatrochemistry, eschewing any tantric content, or reference to transcendent aims or the transmutation of metals.[48] This is broadly the

[46] The section on iatrochemistry in the *Kalyāṇakāraka* breaks this general pattern. However, it is possible that the chapter is a later addition to the work, dating to a later period. The *Rasendramaṅgala*, which contains many medical elements, may be an earlier work than suggested in White 1996: 104. See also Meulenbeld 2000: 717–718 on potential earlier dates for the treatise. This would again change the timeline for the introduction of medicine in Sanskrit alchemical works.

[47] See the sections "Works on rasaśāstra and ratnaśāstra" and "Various works on rasaśāstra and ratnaśāstra" in Meulenbeld 2000: 583–787. The works are ordered by their name rather than their date. The AyurYog timeline of Sanskrit alchemical literature (www.ayuryog.org/alchemytimeline) also presents an overview of the literature. In regard to non-Sanskrit works on Indian alchemy, see Speziale 2019 and 2010: 430–433 for Indian works on alchemy written in Persian, Langermann 2018 for the reception of this literature into Arabic, and Kędzia 2017 for alchemical works of the Tamil Siddha tradition. The latter's history in particular deserves greater attention. The Tamil literature describing mercurial alchemy follows its own trajectory, which we unfortunately cannot present in this volume. It is, however, clear that Tamil and Sanskrit works describe very similar, perhaps interdependent, traditions.

[48] Bhudeb Mookerji's *Rasa Jala Nidhi* of 1926 is somewhat of an exception here, as his book is a compendium of older alchemical works and echoes at least some of their transmutational and transcendent concerns. P. C. Ray's famous 1902 *A History of Hindu Chemistry from the Earliest Times to the Middle of Sixteenth Century* also dates to this period. As the title suggests, this book presented

rasaśāstra that has survived into the present, where it now forms a subdivision of the Indian government-regulated education and practice of Ayurveda and has been integrated into ayurvedic pharmacology.

Making gold and silver, one of the other major aims of alchemists, was not, however, entirely forgotten. A famous experiment, carried out in the 1940s by Krishna Pal Shastri, an ayurvedic pharmacist from Jamnagar, Gujarat, in the laboratory of the Banaras Hindu University, is described as a case of successful transmutation. The experiment was conducted in the presence of a group of ayurvedic scholars and practitioners and the industrialist Birla. It involved pouring mercury into a hollowed-out soap nut, adding several grams of borax, and a grain of a secret powder. The nut was sealed and placed inside a crucible and then roasted. This resulted in a silver-like substance, which later laboratory tests confirmed to be silver. Another, similar experiment produced gold. These experiments are commemorated in marble plaques in the Viśvanāth Temple on the campus of Banaras Hindu University and at the Birla Temple in Delhi.[49]

Evidence for an alchemical practice aimed at producing an immortal body and experiencing liberation in the living body (*jīvanmukti*), on the other hand, is rather more difficult to trace. Although stories of immortal or very ancient perfected beings or Siddhas abound in India, accounts of liberative mercurial practices are less common and seem mostly a matter of hearsay.[50]

In This Volume

Drawing on a selection of the most important Sanskrit alchemical works from the tenth to seventeenth centuries, this volume provides a broad overview of

rasaśāstra works as representing early forms of chemistry, ignoring or rejecting any religious or transmutational aspects of the alchemical tradition. See Mukharji 2016 on Ray's work and Mookerji's critique of it.

[49] For a more detailed description of the experiments, see White 1996: 336–337, and Dash 2003: 23–24. White's description is partly based on the eyewitness account of Yadunandan Upadhyaya, who was present at the Banaras laboratory experiment. Dash translates the Delhi plaques and gives pictures of the Birla Temple in Delhi that houses one of the plaques. The plaque indicates that the Delhi experiment took place in 1942. Another plaque gives 1941 as the date. Both plaques mention Kṛṣṇa Lāla Śarmā as the person who conducted the experiment, and both describe the final product as gold. The second one seems very similar to the experiment recounted by Upadhyaya. See Dash 2003: 23. White 1996: 337 also mentions two newspaper articles from 1968 and 1975, respectively, announcing successful transmutations of mercury into gold.

[50] See White 1996: 350–351 on this subject, and also Newcombe 2017.

premodern Indian alchemy, tracing and explaining alchemical thought as it developed on the Indian subcontinent. It explores the motivations and goals of Indian alchemists and illuminates the theories and methods they developed over time. The ten chapters in this volume reveal the characteristic style of the different works and highlight the key themes that together form the body of knowledge we collectively call "alchemy." The volume also contains one chapter from a different genre of literature, namely a treatise on yoga, chosen for the insight it gives into the spread of alchemical thought and its influence on other disciplines and belief systems.

Each chapter consists of two parts: an introduction and a translation of an excerpt from an alchemical work. The selected excerpts are chosen to reveal the characteristic style of each work and represent the most common themes found in Sanskrit alchemical literature. The introductory sections present the selected alchemical work and its position within Sanskrit alchemical literature, situate the translated chapter within the alchemical work's structure and outlook, provide a discussion of the themes of the chapter and how they represent or are unusual for Sanskrit alchemical thought, and highlight connections to other disciplines such as Ayurveda and Yoga and broader Indic thought in general.

Translation Notes

Each introduction contains a section on the editions used for the translation, as well as notes on translation practices. Technical terms are typically translated into English with the Sanskrit term given in brackets, or in the notes, if a translation is uncertain. We have opted to use common names for plants and other substances, instead of the binomial nomenclature of scientific Latin names. There are two reasons for this. The first is that this renders the text more readable. The underlying assumption here is that the Sanskrit terms used for plants and so on in the original texts were probably widely known and not specialist terms. The second reason is that the Sanskrit terms may denote a group of plants instead of a single, specific plant. The Latin naming system may thus convey the wrong impression of exact plant identification. In those cases in which the identification of a substance is uncertain, we have provided a provisional translation along with an explanatory note. Even common names may not be known to all audiences, though Indian plant materials are becoming somewhat more widely known through the globalization of ayurvedic nutritional supplements and medicines. We have identified plant names using standard Sanskrit-English dictionaries, as well as more specialized plant glossaries and dictionaries, such as *Nadkarni's Indian Materia Medica* (Nadkarni 1954). One of the team of translators, Dr Priyanka Chorge, is a trained ayurvedic practitioner with a Bachelor in Ayurvedic Medicine and Surgery (BAMS) degree. Dr Chorge was often able to provide the common English name of substances as they are understood today for our translations. It may be helpful to readers to know that some websites about Ayurveda provide accessible descriptions of many of the plant materials featured in this volume.

Maps

The maps at the beginning of each chapter show where manuscripts of each respective text are held in South Asia. The maps are based on data from the *New Catalogus Catalogorum*. While they do not tell us the original provenance of each manuscript, they show how many manuscripts of a given treatise survive into the present and provide some insight into how widely a text was distributed. For example, some manuscripts are mostly held in archives in the north of India (*Rasaprakāśasudhākara*, *Rasamañjari*, *Rasendramaṅgala*), while others are found predominantly in the south (*Ānandakanda*),[51] or in Bengal (*Rasendrasārasaṃgraha*). A study of the movements of each manuscript to its current archival holding place could offer fresh insights into the distribution of alchemical thought and perhaps of different lineages of alchemists using one particular text or group of texts. The numbers of catalogued manuscripts vary greatly, with only four known manuscripts of the *Rasakāmadhenu* and over 400 manuscripts (albeit partial ones of separate *khaṇḍas* or books) of the *Rasaratnākara*, about 200 of the *Rasamañjari*, and over 90 of the *Rasaratnasamuccaya*. There are relatively few manuscripts of the oldest of the alchemical works, the *Rasahṛdayatantra* (ca. 20 manuscripts) and *Rasārṇava* (ca. 27). However, as they are extensively quoted in the later texts, their contents survive in different ways.

[51] This volume contains many references to the *Ānandakanda*, but does not have a chapter dedicated to it. However, see the *Ānandakanda* entry in the "Timeline of Sanskrit Alchemical Literature" (http://ayuryog.org/AlchemyTimeline), which shows a map of the surviving manuscripts of the *Ānandakanda*, all of which are located in South India. The timeline also contains maps of other works referenced in this volume.

Chapter 1 by Dagmar Wujastyk introduces the *Rasahṛdayatantra*, the oldest of the Sanskrit alchemical works. The chapter examines the *Rasahṛdayatantra*'s detailed description of the culmination of alchemical practice: the intake of the elixir by the practitioner and his subsequent transformation. Chapter 2, by David G. White, is focused on the second foundational work on *rasaśāstra*, the *Rasārṇava*. Its first chapter lays out the reasons for the use of mercury as an agent for liberation. In its second chapter, the *Rasārṇava* focuses on the practitioners of alchemy: master alchemists/teachers, students, assistants, and female helpers, describing their respective roles, characteristics, and relationships with one another. The chapter outlines a program of initiation for alchemical students that places alchemical practice within a tantric setting, featuring practices such as the fashioning of ritual diagrams (*maṇḍala*s), and the use of spells (*mantra*s, *vidyā*s), seals (*mudrā*s), tantric worship, meditation, protective rites, and so forth.

The third chapter, by Dominik Wujastyk, introduces the *Rasendramaṅgala*, a work whose date is uncertain, although it appears to have been known to Al-Biruni in the early eleventh century. Its fourth chapter describes an exposition of the most famous of the Indian alchemists, Nāgārjuna, on the arts of aurifaction and making elixirs of immortality. The fourth chapter, by Patricia Sauthoff, is a section from the twelfth-/thirteenth-century *Rasendracūḍāmaṇi*. The selected chapter provides a description of the alchemical laboratory (*rasaśālā*) and associated spaces, such as herb gardens and altars for consecrating instruments and materials. This chapter (together with the parallel passages of the *Rasārṇava*) forms the basis for many of the later alchemical works' descriptions of the alchemical workplace. It shows that alchemical work very likely depended on a settled existence and a certain amount of wealth.

Chapter 5, translated and introduced by Keith Cantú, showcases a chapter from the *Rasāyanakhaṇḍa* of the *Rasaratnākara*, one of the longest and most complex of the Sanskrit alchemical treatises. Its chapter on pilgrimage at Srisailam describes what actions an adept (*sādhaka*) should take in order to benefit from the alchemical wonders of Srisailam. The topic of alchemical pilgrimage is unusual for Sanskrit alchemical literature, but nevertheless significant, since it informs pilgrims' ritual activities in the region even today.

In the sixth chapter, Patricia Sauthoff translates a section of the *Rasamañjari* dedicated to topics summarized under the umbrella term of "aphrodisiacs," a mainstay of alchemical literature. The chapter gives recipes for preventing or delaying ejaculating, for increasing the size of the penis, and for curing or causing impotence. It also gives recipes for contraceptives, for the tightening of the vagina, and for the removal of hair. The approach of the alchemists here shows a synthesis of medicine with tantric thought, juxtaposing the use of amulets and charms with the application of herbal and mineral medicines.

In Chapter 7, Priyanka Chorge and Dagmar Wujasyk jointly examine a chapter in the medical section of the *Rasaratnasamuccaya* dedicated to the ayurvedic category of abdominal diseases (Sanskrit: *udararoga*) and their treatment. The *Rasaratnasamuccaya* became the classic work of Indian alchemy and was widely copied. It devotes more than half of its contents to medicine and shows an advanced stage of iatrochemical thought. Its prescriptions follow ayurvedic guidelines, but most of its medicines are mineral-based formulations.

Chapter 8 by Dagmar Wujastyk presents the eighteen alchemical procedures outlined in the *Rasaprakāśasudhākara*. Most of the alchemical works provide a program of alchemical operations that lead to the main product of alchemy, namely the elixir, used for making gold and for attaining immortality and spiritual goals. The chapter discusses the activities of alchemists and examines how their programs of alchemical operations reflect changes in alchemists' aims and methods over time.

In the ninth chapter, Patricia Sauthoff showcases the *Rasakāmadhenu* and its chapter on alchemical equipment. The *Rasakāmadhenu* is a relatively late alchemical work and mostly derivative of older works. However, its exposition on alchemical instruments provides a detailed overview of the apparatuses alchemists used and gives insight into the mechanics of alchemical operations.

Finally, Chapter 10, a translation of a chapter of the *Amṛtasiddhi* by James Mallinson, reflects the reception of alchemical thought into other schools of thought. The *Amṛtasiddhi* is the earliest substantial text on what became *haṭha yoga*. Written in both Sanskrit and Tibetan, the text has a number of Buddhist features. It widely uses alchemical imagery to describe yogic processes such as breathing exercises and their effects on the body and the mind. This imagery then found its way into many of the later *haṭha yoga* texts.

With these ten chapters, the volume brings together texts from a wide range of Indian alchemical thought, each piece serving to expand our understanding of what it meant to practice alchemy on the Indian subcontinent. The volume's translations provide a unique window into this important literature and the Indian alchemical idiom. With its broad selection of examined themes, it also offers the most detailed and comprehensive English-language investigation of Indian alchemy and the beliefs and practices of its practitioners to date.

Outlook

This volume builds on the pioneering work of Ray (1902, 1909), Wujastyk (1984), Roșu (1986, 1992), White (1996), Meulenbeld (2000), Hellwig (2009a, 2009b,2014), and others. The historical study of Indian alchemical traditions is still in its infancy, and much remains to be done, both in terms of philological, historical, and archaeological work and in the ethnographic study of surviving traditions, including the institutionalized forms of rasashastra pharmacology and medicine.

To date, there are no critical editions of Indic alchemical texts. We do have vulgate editions of a number of treatises, most from the early twentieth century. We used several of these for our translations in this volume.[52] Some of these editions are partially or fully available as e-texts on the websites of the DCS, SARIT, or GRETIL. There are very few translations of entire works or even of selected chapters into English or other non-Indic languages. Extant translations are often partial or consist of paraphrases rather than translations. So far, the main academic focus has been on alchemical literature written in Sanskrit, with only a few studies on Tamil sources and Indo-Persian works with medico-alchemical content. To my knowledge, there are no textual or historical studies on alchemical literature composed in other Indic languages, other than Balbir's and Sikdar's surveys of Prakrit texts on alchemy. There is a growing literature on the history of Tibetan Buddhist alchemy, which shows deep connections between Indic and Tibetan medical and alchemical thought, but only few and partial translations of the relevant texts.[53] Further historical connections may exist with Burmese Weikza-Lam traditions of alchemy. Here, there have been some studies on contemporary practices, but no textual studies of older texts. Weikza-Lam practices may represent an intersection between Indian and Chinese forms of alchemy.[54]

[52] The translation by Dominik Wujastyk of a chapter of the *Rasendramaṅgala*, however, is based on his provisional critical edition of the work. See Wujastyk et al. 2022 and also Dominik Wujastyk 2022.

[53] See Simioli 2013, 2015; Czaja 2013; Cantwell 2017.

[54] See Pranke 2014 on Weikza-Lam. For a brief overview of various mercurial medico-alchemical practices across Asia, see Wujastyk 2015.

The history of interactions between Indian and Chinese alchemy is an area of study that deserves more attention, especially since it may aid in explaining why mercury was introduced as the central substance of Indian alchemy.[55]

Indian alchemy can be understood as part of the history of tantric thought. It also has strong links with the Indian medical traditions. Knowledge of mining practices, and of the properties of metals and minerals and their practical applications, formed the basis of alchemical work. However, alchemy's historical links with artisanal traditions, especially metallurgy, are not documented beyond short references in the alchemical treatises. At present, we do not know of literature produced by metallurgical artisans that would provide more detailed evidence for collaborations or knowledge transfer. Here, texts written in Indic vernaculars may provide key findings.

There are many ways in which the academic study of the history of Indian alchemy can be enriched. We trust that this volume lays a secure foundation for further study.

Bibliography

Balbir, Nalini. 1990. "Scènes d'alchimie dans la littérature jaina." *Journal of the European Ayurvedic Society*, 1: 149–164.

Balbir, Nalini. 1992. "La fascination jaina pour l'alchimie." *Journal of the European Āyurvedic Society* 2: 134–150.

Baldissera, Fabrizia. 2015. "Traces of Early Alchemy in India. Rasāyana in Some Kāvya and Kathā Texts with a Note on Chinese Alchemy." *AION* 37: 13–34.

Beal, Samuel. 1884. *Si-Yu-Ki. Buddhist Records of the Western World. Translated form the Chinese of Hiuen Tsiang (A.D. 629)*. Volume 2. London: Trübner & Co.

Birch, Jason. 2024. *The Amaraugha and Amaraughaprabodha of Gorakṣanātha: The Genesis of Haṭha and Rājayoga. A Critical Edition and Annotated Translation by Jason Birch*. Pondicherry: Institut Français de Pondichéry.

Cantwell, Catherine. 2017. "Reflections on Rasāyana, Bcud len and Related Practices in Nyingma (Rnying ma) Tantric Ritual." *History of Science in South Asia* 5 (2): 181–203.

Chojnacki, Christine, and Hampa Nagarajaiah, eds. 2018. *Uddyotanasūri's Kuvalayamālā. A Jain Novel From 779 AD*. Translated from the French by Alexander Reynolds and largely revised by the Author. Volume 2. Bangalore: Sapna Book House.

Czaja, Olaf. 2013. "On the History of Refining Mercury in Tibetan Medicine." *Asian Medicine: Tradition and Modernity* 8 (1): 75–105.

Dash, Bhagwan. 2003. *Alchemy and Metallic Medicines in Ayurveda*. New Delhi: Concept Publishing Company.

DCS. Digital Corpus of Sanskrit. http://www.sanskrit-linguistics.org/dcs/.

Fenner, Edward Todd. 1979. "Rasayana Siddhi: Medicine and Alchemy in the Buddhist Tantras." PhD diss., University of Wisconsin–Madison.

[55] See White 1996: 61–63 and Needham et al. 1976: 160. The study by Yang 2023 points to an entangled history of Indian and Chinese breathing and physical movement practices in which alchemical terminology is used.

GRETIL. Göttingen Register of Electronic Texts in Indian Languages. https://gretil.sub.uni-goettingen.de/gretil.html.

Hellwig, Oliver. 2009a. "A Chronometric Approach to Indian Alchemical Literature." *Literary and Linguistic Computing* 24 (4): 373–383. https://doi.org/10.1093/llc/fqn043.

Hellwig, Oliver. 2009b. *Wörterbuch der mittelalterlichen indischen Alchemie.* Groningen: Barkhuis.

Hellwig, Oliver. 2014. "Materialgruppen in der indischen Alchemie." *Zeitschrift der Deutschen Morgenländischen Gesellschaft* 164 (2): 451–468.

Kędzia, Ilona Barbara. 2017. "Mastering Deathlessness: Some Remarks on Kaṛpam Preparations in the Medico-Alchemical Literature of the Tamil Siddhas." *History of Science in South Asia* 5 (2): 121–142. https://doi.org/10.18732/hssa.v5i2.16.

Langermann, Y. Tzvi. 2018. "The Chapter on *Rasāyana* (Medications for Rejuvenation) in *Mi'rāj al-du'ā'*, a Shi'ite Text from the 12th/18th Century." In *Intellectual History of the Islamicate World* 6 (1–2): 144–183. https://doi.org/10.1163/2212943X-00601010.

Maas, Philipp A. 2017. "Rasāyana in Classical Yoga and Āyurveda." *History of Science in South Asia* 5 (2): 66–84. https://doi.org/10.18732/hssa.v5i2.32.

Mallinson, James. 2019. "Kālavañcana in the Konkan: How a Vajrayāna Haṭhayoga Tradition Cheated Buddhism's Death in India." *Religions* 10 (4): 273. https://doi.org/10.3390/rel10040273.

Mallinson, James. 2020. "The Amṛtasiddhi: Haṭhayoga's Tantric Buddhist Source Text." In *Śaivism and the Tantric Traditions*, edited by Dominic Goodall, Shaman Hatley, Harunaga Isaacson, and Srilata Raman, 409–425. Leiden: Brill. https://doi.org/10.1163/9789004432802_019.

Mallinson, James, and Péter-Dániel Szántó, eds. 2022. *Amṛtasiddhi*. Hatha Yoga Series of the Collection Indologie of the Institut Français d'Indologie/École française d'Extrême-Orient, Pondicherry. Pondicherry: Institut Français d'Indologie/École française d'Extrême-Orient.

Meulenbeld, Gerrit Jan. 2000. *A History of Indian Medical Literature.* Volume 2A. Groningen: Egbert Forsten.

Mookerjee, Bhudeb. 1926–1938. *Rasa-Jala-Nidhi or Ocean of Indian Chemistry, Medicine and Alchemy. Compiled in Sanskrit by Bhudeb Mookerjee With English Translation by the Author.* Volumes 1–5. Calcutta: Nababibhakar Press.

Mukharji, Projit. 2016. "Parachemistries: Colonial Chemopolitics in a Zone of Contest." *History of Science* 54 (4): 362–382.

Nadkarni, A. K. 1954. *Dr. K. M. Nadkarni's Indian Materia Medica, with Ayurvedic, Unani-tibbi, Siddha, Allopathic, Homeopathic, Naturopathic and Home Remedies, Appendices and Notes.* 2 vols. Bombay: Popular Prakashan.

Needham, Joseph, Ho Ping-Yü, and Lu Gwei-Djen. 1976. *Science and Civilization in China*, Volume 5: *Chemistry and Chemical Technology*, Part 3: *Spagyrical Discovery and Invention: Historical Survey, From Cinnabar Elixirs to Synthetic Insulin.* Cambridge: Cambridge University Press.

Newcombe, Suzanne. 2017. "Yogis, Ayurveda, and Kayakalpa: The Rejuvenation of Pandit Malaviya." *History of Science in South Asia* 5 (2): 85–120. https://doi.org/10.18732/hssa.v5i2.29.

Pranke, Patrick. 2014. "On Saints and Wizards: Ideals of Human Perfection and Power in Contemporary Burmese Buddhism." In *Champions of Buddhism: Weikza Cults in Contemporary Burma*, edited by Brac de la Perrière, Bénédicte, Rozenberg, Guillaume, and Turner, Alicia, 3–31. Singapore: National University of Singapore Press.

Principe, Lawrence M. 2013. *The Secrets of Alchemy.* Chicago: University of Chicago Press.

Ray, Prafulla Chandra. 1902, 1909. *A History of Hindu Chemistry from the Earliest Times to the Middle of the 16th Century A.D.* 2 vols. Calcutta: Prithwis Chandra Ray.

Rongxi, Li. 1996. *The Great Tang Dynasty Record of the Western Regions.* Translated by Li Rongxi. Taishō 51, no. 2087. Moraga, CA: BDK America, Inc.

Roşu, Arion. 1986. "*Mantra* et *Yantra* dans la médecine et l'alchimie indiennes." *Journal Asiatique* 274 (3–4): 205–268.

Roşu, Arion. 1992. "Alchemy and Sacred Geography in the Medieval Deccan." *Journal of the European Āyurvedic Society* 2 (1): 151–157.

Sachau, Edward C. 1910. *Al-Beruni's India: An Account of the Religion, Philosophy, Literature, Geography, Chronology, Astronomy, Customs, Laws, and Astrology of India about 1030AD. An English Language Edition with Notes and Indices by Dr. Edward C. Sachau in Two Volumes.* London: Kegan Paul, Trench, Trubner & Co. Ltd.

SARIT. Search and Retrieval of Indic Texts. https://sarit.indology.info/.

Sikdar, J. C. 1980. "Jaina Alchemy." *Indian Journal of History of Science* 15 (1): 6–17.

Simioli, Carmen. 2013. "Alchemical Gold and the Pursuit of the Alchemical Elixir." *Asian Medicine: Tradition and Modernity* 8 (1): 43–77.

Simioli, Carmen. 2015. "The King of Essences: Mercury within Tibetan Medico-Alchemical Traditions." In *Elisir Mercuriali e Immortalità Capitoli per Una Storia dell'Alchimia nell'Antica Eurasia*, edited by Giacomella Orofino et al., 35–54. Pisa and Rome: Fabbrizio Serra Editore.

Speziale, Fabrizio. 2006. "De zeven vrienden. Een Indo-Perzische verhandeling over alchemie." In *Bronnen van kennis. Wetenschap, kunst en cultuur in de collecties van de Leidse Universiteitsbibliotheek*, edited by P. Hoftijzer, K. van Ommen, G. Warnar, and J. J. Witkam, 23–31. Leiden: Primavera Pers.

Speziale, Fabrizio. 2010. "Les traités persans sur les sciences indiennes: médecine, zoologie, alchimie." In Hermann, Denis & Fabrizio Speziale (eds.) *Muslim Cultures in the Indo-Iranian World during the Early-Modern and Modern Periods.* Berlin: Klaus Schwarz Verlag.

Speziale, Fabrizio. 2019. "Rasāyana and Rasaśāstra in the Persian Medical Culture of South Asia." *History of Science in South Asia* 7: 1–41. https://doi.org/10.18732/hssa.v7i0.40.

Stapleton, H. E., and R. F. Azo. 1905. "Alchemical Equipment in the Eleventh Century, A.D." *Memoirs of the Asiatic Society of Bengal* 1 (4): 47–70.

Upadhyaya, Jagannatha. 1986. *Vimalaprabhāṭīkā of Kalki Śrī Puṇḍarīka on Śrī Laghukālacakratantrarāja by Śrī Mañjuśrīyaśa.* Volume 1. Sarnath, Varanasi: Central Institute of Higher Tibetan Studies.

Upādhye, Ādinātha Neminātha. 1959. *Uddyotana -Sūri's Kuvalayamālā (A Unique Campū in Prākrit). Critically Edited from Rare Mss Material for the First Time with an Introduction, Various Readings, Notes etc. and Ratnaprabha- Sūri's Kuvalayamālā-Kathā (A Stylistic Digest of the Above in Sanskrit) Critically Edited with Various Readings etc. by Professor A. N. Upadhye. Part I. Kuvalayamālā Prākrit-Text and Various Readings.* Bombay: Singhi Jain Śāstra Śikṣāpīṭha, Bharatiya Vidya Bhavan

Waley, Arthur. 1932. "References to Alchemy in Buddhist Scriptures." *Bulletin of the School of Oriental and African Studies* 6 (4): 1102–1103. https://doi.org/10.1017/S0041977X00123535.

White, David Gordon. 1995. "The Ocean of Mercury: An Eleventh-Century Alchemical Text." In *Religions of India in Practice*, edited by Donald S. Lopez, 281–287. Princeton: Princeton University Press.

White, David Gordon. 1996. *The Alchemical Body: Siddha Traditions in Medieval India.* Chicago: University of Chicago Press.

Wujastyk, Dagmar. 2013. "Perfect Medicine: Mercury in Sanskrit Medical Literature." *Asian Medicine* 8 (3): 15–40.

Wujastyk, Dagmar. 2015. "Histories of Mercury in Medicine across Asia and Beyond." *Asiatische Studien—Études Asiatiques* 69 (4): 819–830.

Wujastyk, Dagmar. 2017. "Acts of Improvement: On the Use of Tonics and Elixirs in Sanskrit Medical and Alchemical Literature." *History of Science in South Asia* 5 (2): 1–36. https://doi.org/10.18732/hssa.v5i2.26.

Wujastyk, Dagmar. 2019. "Iron Tonics: Tracing the Development from Classical to Iatrochemical Formulations in Ayurveda." *HIMALAYA* 39 (1): 74–90. https://digitalcommons.macalester.edu/himalaya/vol39/iss1/11.

Wujastyk, Dagmar. 2023. "The Forge and the Crucible. Images of Alchemical Apparatuses on Manuscripts of the Rasendramaṅgala." In *Suhṛdayasaṃhitā. A Compendium of Studies on South Asian Culture, Philosophy, and Religion Dedicated to Dominik Wujastyk*, edited by Philipp A. Maas and Anthony Cerulli, 127–164. Halle a.d.S.: Universitätsverlag Halle-Wittenberg (Studia Indologica Universitatis Halensis).

Wujastyk, Dominik. 1984. "An Alchemical Ghost: The Rasaratnâkara by Nâgârjuna." *Ambix* 31 (2): 70–83.

Wujastyk, Dominik. 2022. "The Surviving MSS of the Rasendramaṅgala of Nāgārjuna Siddha." https://doi.org/10.5281/zenodo.7191411.

Wujastyk, Dominik, Chakraborty, Deepro, Bhatt, Harshal, Lele, Vanadana and Allred, Jane. 2022. "The Rasendramaṅgala: Manuscript Transcriptions." *Zenodo*. https://doi.org/10.5281/zenodo.6471657.

Yang, Dolly. 2023. "Knowledge Transfer of Bodily Practices Between China and India in the Mediaeval World." *Journal of Yoga Studies* 4: 413–440.

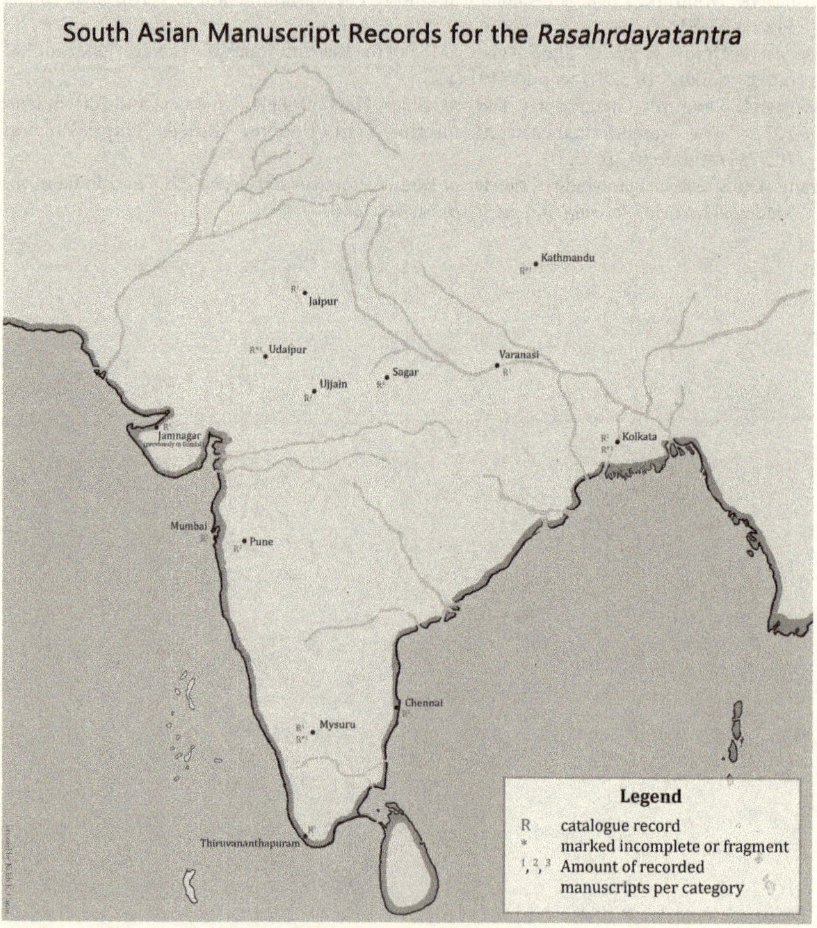

Map 1.1 South Asian manuscript records for the *Rasahṛdayatantra*.

1

The Culmination of Alchemical Practice

Chapter 19 of the *Rasahṛdayatantra*

Dagmar Wujastyk

When Indian alchemists started composing their works, they defined the goals of alchemy and the parameters of their discipline. They also devised systematic programs for achieving their goals. In all works, the key to achieving the alchemical endeavor lies in the use of one substance in particular: mercury. Combined with other substances, and subjected to various alchemical operations, mercury is used to make the ultimate alchemical product, the mercurial elixir. While there is general agreement on the central importance of mercury, the Sanskrit alchemical works vary in many of their specifications about what other materials should be applied in the alchemical processes, what alchemical operations should be employed, and in what order these should be executed. The early works describe two broad aims: first, the transmutation of metals and gems, and second, the transformation of the bodily and/or spiritual condition of the practitioner or consumer. However, in regard to the latter, even the earliest works display substantial differences in their explanations of what kind of transformation they hope to achieve with their alchemical products. Indeed, even a single treatise can list a wide range of desired effects. Projected outcomes range from improved health, revitalization of body and mind, and increased longevity to the attainment of special powers (*siddhi*) and immortality, and beyond that, to the complete transcendence of the human condition to enter a god-like state or merge with an absolute reality.

In this chapter, we will examine the final stage of the alchemical program of the *Rasahṛdayatantra* (*The Book of the Heart of Mercury*) by the monk Govinda, one of the oldest Sanskrit alchemical works transmitted to us. Dating to the tenth or eleventh century CE, it is a relatively short treatise of

Dagmar Wujastyk, *The Culmination of Alchemical Practice* In: *Indian Alchemy*. Edited by: Dagmar Wujastyk,
Oxford University Press. © Oxford University Press 2025. DOI: 10.1093/9780197805831.003.0002

506 verses, arranged in nineteen chapters.[1] The work delineates a concise and systematic program of eighteen alchemical operations (*rasakarman*; *saṃskāra*[2]): sixteen for making the elixir, and a further two sets of procedures for the final processes of applying the elixir to metals and the practitioner/consumer. Its list of alchemical operations broadly informs the structure of the work. Eighteen of the nineteen chapters are dedicated to them, loosely following the order in which each operation is to be performed.[3] The first chapter prepares the ground for the more technical, prescriptive part of the work. It introduces the reader to the subject of mercury as an all-powerful substance through which one can conquer all disease, attain special powers, and reach final emancipation (*mukti*). The author explains that only with a stable body (*sthiradeha*) will one be able to attain the highest goal of spiritual liberation. According to Govinda, the purpose of alchemy is therefore to prolong life and to stabilize both body and mind to enable longer spiritual study and practice, which in turn will lead to the ultimate goal of attaining absorption into the Absolute.[4] Mercury is the substance with which this can be achieved. The reason for this lies in mercury's special characteristics of imperishability and its power to absorb other materials, which make it superior to other substances: "No potion of roots or metals that is itself unstable, can be burnt, rotted or dried up is capable of producing its (i.e., the body's) stability. Woods and herbs are absorbed in lead, lead in tin, and tin in turn in copper, copper in silver, silver in gold and gold in mercury."[5] However, first,

[1] On the question of the dating of the *Rasahṛdayatantra*, see Meulenbeld 2000: 621. Compare its length with that of the roughly contemporaneous *Rasārṇava* (ca. 2,300 verses), or later important alchemical works, like the *Rasaratnākara* (more than 8,000 verses), the *Rasendracūḍāmaṇi* (ca. 1,400 verses), the *Rasakāmadhenu* (more than 5,000 verses), or the *Rasaratnasamuccaya* (ca. 3,870 verses). Verse counts can vary depending on the editions. The ones cited here follow the count in Meulenbeld 2000.

[2] The *Rasahṛdayatantra* only uses the term *saṃskāra* once, in the final verse of the first chapter. See Chapter 8 in this volume for a more detailed discussion of the topic of *saṃskāra*.

[3] Even though the number of chapters parallels the number of alchemical procedures (with the first chapter serving as an introduction), the work does not dedicate a full chapter to each of them. Instead, some chapters combine descriptions of procedures for mercury with descriptions of cleansing or potentiating procedures for other materials. For example, chapter 3 describes the processing of crude mica for its combination with mercury, but also describes various methods of "feeding" (*cāraṇa*) it to mercury. Chapter 4 describes the extraction of the essence of mica and its fixation (*bandhana*), followed again by its combination with mercury.

[4] Terms used for this are *mukti* (verses 3, 10), *brahmatva* (verse 2), *brahmapada* (verse 15), and *brahmabhāva* (verse 26). Liberation in a living body (*jīvanmukti*) is also mentioned (verses 32 and 33).

[5] *Rasahṛdayatantra* 1.11–12. *tatsthairyaṃ na samarthaṃ rasāyanaṃ kimapi mūlalohādi* |
svayam asthirasvabhāvaṃ dāhyaṃ kledyaṃ ca śoṣyaṃ ca ||11||
kāṣṭhauṣadhyo nāge nāgaṃ vaṅge vaṅgamapi līyate śulve |
śulvaṃ tāre tāraṃ kanake kanakaṃ ca līyate sūte ||12||.

mercury (and also every other material used) must go through a series of alchemical operations that will render it fit for purpose. Through these, mercury will be cleansed of impurities and negative characteristics. Its innate good properties will be enhanced and its power heightened, until it finally attains the right condition needed for its purpose as a transmutative agent and an elixir of immortality and transcendence.

The second chapter provides a list of the eighteen alchemical operations for this purpose and gives brief instructions for the first eight procedures, which are collectively categorized as "*śodhana*," that is, the purification, perfecting, or readying of mercury.[6] The second set of procedures is described in more detail in the chapters that follow. The *Rasahṛdayatantra*'s list of operations comprises: (1) steaming (*svedana*), (2) trituration (*mardana*), (3) thickening (*mūrchā*), (4) raising (*utthāpana*), (5) causing to fall (i.e., evaporation and condensation) (*pātana*), (6) reviving (*nirodha*), (7) fixing (*niyama*), (8) stimulating (*dīpana*), (9) measuring the morsel of mica (*gaganagrāsapramāṇa*), and (10) feeding it to the mercury (*cāraṇa*), (11) enclosed liquefaction (*garbhadruti*), (12) external liquefaction (*bahyadruti*), (13) "digestion" (i.e., amalgamation; *jāraṇa*), (14) coloring mercury (*rasarāga*), (15) "setting in motion" or potentiation (*sāraṇa*), (16) advancing (*krāmaṇa*), (17) transmutation (*vedha*), and finally (18) ingestion (*bhakṣaṇa*). Some of the procedures receive more attention—or at least, more descriptive detail—in the *Rasahṛdayatantra* than others. The longest chapters are those dedicated to the two final steps: the transmutation of metals, and the transmutation of the practitioner.

The last set of procedures, namely those pertaining to the consumption of the alchemical elixir and the subsequent transformation of the practitioner, are described at length in the nineteenth and final chapter of the *Rasahṛdayatantra*. In its list of *saṃskāra*, or *rasakarman*, in chapter 2, the *Rasahṛdayatantra* calls this final set of procedures *bhakṣana*, that is, ingestion. However, chapter 19, verse 1 uses a different term: *rasāyana*, which I have translated as "elixir regimen" here.[7] This term is first found in Sanskrit

[6] This categorization is, however, only found in the chapter colophon, which reads "*iti parama haṃsaparivrājakācāryaśrīmadgovindabhagavatpādaviracite rasahṛdayatantre* ***rasaśodhanātmako*** *dvitīyo'vabodhaḥ //*."

[7] Verse 1 reads: *iti rasarājasya vidhau vedhavidhānaṃ prasaṃgataḥ proktam / adhunā proktānapi vakṣyāmi* ***rasāyane*** *yogān // 1 //*. Common alternative terms for this stage of the alchemical process found in other alchemical works are *sevana*, *bhakṣana*, *śarīrayoga*, and *dehavedha* ("the transmutation of the body"), see White 1996: 265. The translation I have chosen here, "elixir regimen," is not a literal translation of the compound term *rasāyana*, but a representation of what the term refers to in the alchemical context. On the difficulties of translating the term *rasāyana*, see Maas 2017: 75–79, and Dagmar Wujastyk 2017:1–2.

medical literature, where it denotes one of eight traditional subject areas of ayurvedic knowledge. Most premodern Sanskrit medical works dedicate a section to *rasāyana* therapy, vitalization treatments that are meant to counter the effects of ageing and to prolong life. Medical *rasāyana* consists of a series of therapeutic interventions that begin with various purgative treatments for inner cleansing, followed by a period of recovery and a specific dietary regimen and concluded by the main medical application: the intake of a tonic over a period of time. The expected results include better general health, rejuvenation, and a prolonged lifespan. In a few instances, the promised results are bolder, and consumers of *rasāyana* tonics are promised an indefinite extension of lifespan and an indestructible body.

The alchemical *rasāyana*, or elixir regimen, is clearly modeled on the older therapeutic procedures of Ayurveda. It differs, however, in its focus on mercury, a substance that is not used for vitalization therapy in the early medical works, and that only becomes a minor ingredient in later ones.[8] And while recipes for ayurvedic *rasāyana* tonics can be complex, none of them require quite as complicated preparation methods as are standard for the mercurial elixirs of alchemy.

The *Rasahṛdayantantra*'s chapter on elixir regimen outlines a series of activities and recipes, broadly consisting of the preliminary treatment of cleansing the body (*śodhana*) through purgation, sweating, and emesis—a process called "the preparation of the body" (*kṣetrīkaraṇa*) involving the intake of herbal and mercurial elixirs (in particular a mercury preparation called *āroṭa*)—and the intake of a mercurial elixir that triggers the ultimate transformation of the practitioner's body.[9] However, these stages are not well distinguished from one another, and it is not always clear whether the described herbal and mercurial elixirs are meant to be taken in succession or whether they are alternatives to one another. The transformation of the practitioner is described as a lengthy process that is drawn out over a period of time, rather than as something that happens in an instant after taking an elixir. Some of the transformations seem to already take place as part of the preparatory treatments. These might be considered successive stages in the course of a transformation that culminates in liberation, but they may also

[8] On the similarities and differences between the vitalization therapies described in ayurvedic works and the last stage of the alchemical process, see Dagmar Wujastyk 2017.

[9] White 1996: 266 also proposes "making (oneself master of) the field" for *kṣetrīkaraṇa*. Hellwig's dictionary does not have a separate entry for *kṣetrīkaraṇa*, but explains it as "Vorbereitung des Körpers auf den *dehavedha*"—"preparation of the body for *dehavedha*" (Hellwig 2009: 184).

represent independent and equal options. Indeed, the *Rasahṛdayatantra* seems to advocate different pathways for both its course of elixir regimen and the potential goals achieved with it.

The *Rasahṛdayatantra*'s *rasāyana* chapter shares a number of features with the *rasāyana/dehavedha/sevana* chapters or passages of later alchemical works, in terms of both general structure of the treatment process and the required conditions for it. All early alchemical works describing an elixir regimen prescribe cleansing procedures that precede the intake of the elixirs, though there are significant differences in how much detail is given on how to proceed. Most of these alchemical works also include the concept of "the preparation of the body" (*kṣetrīkaraṇa*) in their presentation of an elixir regimen.[10]

Cleansing and Preparing the Body

The internal cleansing of the body through purgation and so on is very similar to what ayurvedic works prescribe for preparing for ayurvedic vitalization therapy and aims at ridding the body of excess waste or humoral substances. The *Rasahṛdayatantra*'s list of substances used for internal cleansing corresponds closely with the drugs used in preparation for vitalization therapy in early ayurvedic works, though it uses different names for them.[11] Later alchemical works follow the *Rasahṛdayatantra*'s format and prescribe similar preparations.[12]

[10] The *Rasaratnasamuccaya* does not mention *kṣetrīkaraṇa* in its *rasāyana* chapter (chapter 26) but makes mention of it in 11.66 in the context of the purified mercury preparation called "*āroṭa*" as the most suitable "preparation of the body." Similarly, the *Rasārṇava* (18.10–11) associates *kṣetrīkaraṇa* with the intake of *āroṭa*, though also with a number of other mercurial preparations (see 18.16, 18–19, and 21). The *Rasaratnākara*'s *Rasāyanakhaṇḍa* makes a fairly clear distinction between cleansing procedures and *kṣetrīkaraṇa*, as it describes cleansing procedures in chapter 1.4–8, and *kṣetrīkaraṇa* (with mica) in chapter 2.2–3.

[11] See, for example, *Carakasaṃhitā Cikitsāsthāna* 1.1.25–28, which lists *harītakī* (chebulic myrobalan), *saindhava* (rock salt), *āmalaka* (emblic myrobalan), *guḍa* (jaggery), *vacā* (sweet flag), *viḍaṅga* (false black pepper), *rajanī* (turmeric), *pippalī* (long black pepper), and *viśvabheṣaja* (dried ginger) for cleansing treatments. The *Rasahṛdayatantra* uses the term *pathyā* instead of *harītakī* for chebulic myrobalan, *dhātrī* instead of *āmalaka* for emblic myrobalan, *śuṇṭhī* instead of *viṣvabheṣaja* for dried ginger, and also adds *marica* (black pepper). Compare also *Aṣṭāṅgahṛdayasaṃhitā Uttarasthāna* 39.21 and *Aṣṭāṅgasaṃgraha Uttarasthāna* 49.14, which list *harītakī* (chebulic myrobalan), *āmalaka* (emblic myrobalan), *saindhava* (rock salt), *nāgara* (dried ginger), *vacā* (sweet flag), *haridrā* (turmeric), *pippalī* (long black pepper), *vella* (false black pepper), and *guḍa* (jaggery) as the substances used for cleansing. The same list is found in *Siddhayoga* 69.4–5.

[12] *Rasahṛdayatantra* 19.5–7, for example, is echoed in *Rasārṇava* 18.3–7, while a similar, though not identical, cleansing regimen is found in *Rasaratnākara Rasāyanakhaṇḍa* 1.4–6. The *Ānandakanda*'s detailed description of preparing the body in 1.6 mentions similar herbal substances

Both alchemical and medical works recommend a recovery period after internal cleansing with a simple diet of grain and clarified butter. In the medical context, this is followed by the intake of the vitalization tonic. In the alchemical context, the *kṣetrīkaraṇa* process follows.

The *Rasahṛdayatantra* gives a series of recipes with deodar cedar, which seem to be used for *kṣetrīkaraṇa.*[13] Their effects are notable and include cleansing out the abdomen, effecting an increase in beauty and mental vigor, allaying all disease, and becoming immortal and endowed with great vital power. However, treatment with deodar can also have less dramatic outcomes, such as the cure of eye disorders, colds (*pīnasa*), skin diseases (*kuṣṭha*), or generally of diseases caused by each of the humoral substances.[14]

The next section of the *Rasahṛdayatantra* is concerned with various mica (*ghana, abhra, gagana*) preparations and mica and iron compounds. Mica is presented as a substance that prolongs life. For example, verse 23 speaks of a person living for one hundred years. However, the treatise also warns of the detrimental effects of the wrong kind of or badly prepared mica, which would cause decrepitude and death. Mica with iron is said to remove obesity, various eye diseases, swellings, pain in the ears and nose, piles, haemorrhoids, urinary disorders, and grey hair.[15] Taken for a year, it wards off death and ageing.

The focus on mica may be one of the features of the alchemical works' elixir regimens that influenced ayurvedic vitalization therapies. The older ayurvedic works do not mention mica as an ingredient in their *rasāyana* recipes.[16] The first time mica occurs in a medical vitalization treatment is in the eleventh-century *Cakradatta*, where it is used for its "immortality-essence iron" (*amṛtasāralauha*) recipe.[17] Cakrapāṇidatta, the author of the *Cakradatta*, ascribes this recipe to the alchemist Nāgārjuna.

as the *Rasahṛdayatantra*. In some cases where the *Rasahṛdayatantra*'s list of substances diverges from what is found in the medical works, we find correspondences in the alchemical works. For example, the *Rasārṇava*, the *Rasaratnākara*, and the *Ānandakanda* also refer to the use of screw pine, and the *Rasārṇava* includes hellebore, while none of the examined medical works mention either substance. See *Rasārṇava* 18.3 and *Rasaratnākara Rasāyanakhaṇḍa* 1.5 on the use of screw pine, and *Rasārṇava* 18.5 on the use of hellebore.

[13] Deodar also plays a role in other alchemical works in the context of *kṣetrīkaraṇa*. In the *Rasārṇava* (18.9–10), the protracted use of deodar cedar is followed by the intake of processed mercury. In the *Ānandakanda*, deodar is featured in 1.6.25–26 as the last of the preparatory preparations before *rasāyana* with processed mercury (*āroṭa*) begins.

[14] On disease causation through humoral substances (wind, bile, and phlegm) in ayurvedic medicine, see Dominik Wujastyk 2003: xvii–xviii, and for more detail, Mazars 2006: 31–35.

[15] See *Rasahṛdayatantra* 19.32.

[16] See Dagmar Wujastyk 2017: 22–23.

[17] *Cakradatta* 66.92–97.

Safety and Efficacy

The *Rasahṛdayatantra* stresses the importance of assuring the efficacy of mercurials through proper preparation, but also highlights the issue of safety, noting that mercury would damage the body of someone who had not properly prepared for treatment: "Mercury will not penetrate the body of one who undertakes an elixir regimen without having prepared his body. The mercury will cause damage to all parts of the body."[18] Thus, not only does mercury have to undergo treatment to become suitable for use (see below), the practitioner also needs to prepare in order to become a suitable recipient. One of the better-known aphorisms of the Indian alchemical tradition is a statement from the *Rasārṇava* (17.165–166), which posits a correspondence between metals and the human body ("*yathā lohe tathā dehe*"—"as in metals, thus in the body"). Because of their parallel nature, mercury can be applied to metals first, and its effect on the metals will indicate whether it has become suitable for use on the human body. However, here, it is mercury and the human body that share characteristics (namely, impurities) and thus must be treated in similar ways.

Rasahṛdayatantra 19.37–58 outlines the proper use of mercurials. This passage discusses what kind of mercury should be used for elixir regimen, namely mercury that has undergone processing and has amalgamated with other metals and minerals. It also clarifies what kind of mercury ought not to be used, namely mercury amalgamated with poison, tin, or lead. It outlines what happens if such mercury is eaten, namely that the consumer would develop skin diseases and trembling.

The chapter also gives a brief overview of how the processes of mercury "digesting" other substances relate to the dosage of the end product. It specifies the regimen and diet during the period the mercurial is taken, what happens if these rules are not followed, the symptoms of not being able to digest mercury (faintness, sleepiness, fever, burning sensations, sharp abdominal pain, apathy, and destruction of the body), and how to treat the arising disorders.[19]

[18] *Rasahṛdayatantra* 19.8: *akṛtakṣetrīkaraṇe rasāyanaṃ yo naraḥ prayuñjīta / tasya krāmati na rasaḥ sa rasaḥ sarvāṅgadoṣakṛdbhavati //.*

[19] Notably, the subject of mercury poisoning is absent from the early ayurvedic works that prescribe medicines with mercury. On this subject, see Dagmar Wujastyk 2013: 31–35.

Outcomes

Several passages in chapter 19 refer to the expected outcomes of the completed elixir regimen process. Verse 49 seems to describe a mercurial-enhanced meditation practice that culminates in arriving at a state called *rasānanda*, "bliss of mercury": "He has become absorbed in the highest being. With a stilled mind, he has achieved equanimity. Controlling his breathing, having mastered the three goals, he has become completely content in the bliss of mercury."[20] The effects of elixir regimen described in verses 61–64 include producing beautiful offspring, enhanced virility, rejuvenation, increased mental power and bodily strength, and prolonged lifespan. There is a sense of gradual improvement over time that corresponds to the prolonged intake of the elixir. And then, finally, the practitioner seems to overcome the human condition altogether, leaving behind "troubles, ageing and death," roaming through the air and becoming god-like: "He will also become a giver and a creator here in the triad of worlds, like the Lotus-born; one who maintains, like Vishnu, and a destroyer, like Rudra."[21] Notably, liberation (*mukti*) does not feature in this description of outcomes or indeed anywhere in chapter 19, despite being listed as an effect of mercury in chapter 1.

Chapter 19 does not end here, but continues with recipes for several pills: a mercurial pill called "immortal beauty" (*amarasundarī*) for protection against weapons and diseases; a pill called "raiser of the dead" (*mṛtasaṃjīvanī*) against injuries, fear, grief, disease, ageing, unhappiness, and indeed, for raising the dead; three kinds of "adamantine pill" (*vajriṇīguṭikā*), said to confer the strength of nine elephants, make the body indestructible and free it from death, ageing, and disease;[22] and one "roaming the sky pill" (*khecarīguṭikā*), through which one would become "highly revered by gods, demons and perfected beings, beginning with Indra."[23] All of these are either placed in the mouth or worn as an amulet.

Many of the projected outcomes of the *Rasahṛdayatantra*'s elixir regimen, such as longevity, strength, beauty, mental vigor, getting rid of wrinkles and grey hair, freedom from disease, and great vital power, echo the effects

[20] *parame brahmaṇi līnaḥ praśāntacittaḥ samatvam āpannaḥ / āśvāsayan trivargaṃ vijitya rasānandaparitṛptaḥ //.*

[21] *Rasahṛdayatantra* 19.64 *dātā bhuvanatritaye sraṣṭā so'pīha padmayonir iva / bhartā viṣṇur iva syāt saṃhartā rudravad bhavati //.*

[22] *Rasahṛdayatantra* 19.74: *eṣā mukhakuharagatā kurute navanāgatulyabalam / tad vapur api durbhedyaṃ mṛtyujarāroganirmuktam //.*

[23] *Rasahṛdayatantra* 19.76cd: *devāsurasiddhagaṇaiḥ pūjyatamo bhavati cendrādyaiḥ //.*

of *rasāyana* therapy in the medical works. The more specific diseases mentioned in the *Rasahṛdayatantra* include eye disorders, including classes of eye disorders called *paṭala*, *kāca*, and *timira*; colds (*pīnasa*); serious skin diseases (*kuṣṭha*); diseases arising through disorders of the humoral substances wind, bile, and phlegm (*vāta*, *pitta*, *kapha*); swellings (*arbuda*); ear and nose pain; piles (*arśa*); anal fistula (*bhagandara*); and urinary (*meha*) and spleen (*plīha*) disorders. All of these are also described in the medical works' *rasāyana* chapters, with the exception of the *kāca* and *paṭala* classes of eye disorders. However, generally speaking, the alleviation of diseases seems a minor concern in the *Rasahṛdayatantra*'s depiction of elixir treatment, and this is also true for the other early alchemical works.[24]

As we have seen, the *Rasahṛdayatantra* associates the preparation of the body (*kṣetrīkaraṇa*) with a wide range of effects, from overcoming disease to attaining immortality in a physical body. And even further: verse 36 claims that the described *kṣetrīkaraṇa* preparations, taken for half a year to a year, eventually lead to the attainment of ultimate bliss (*niḥśreyasasiddhi*). This suggests that a substance used for *kṣetrīkaraṇa* and one used following *kṣetrīkaraṇa* may be one and the same, the difference lying only in the amount of time they are taken. In any case, it is notable that the attainment of immortality is considered part of preparing the body, suggesting that it is not the final aim in the alchemist's endeavor. The step beyond immortality would appear to be the attainment of a god-like status, as described in verses 63–64. However, the recipes for the various pills at the end of the chapter point to the possibility of multiple aims of alchemists rather than one single final aim.

Finally, it should be noted that it is not entirely clear who is undertaking elixir treatment, that is, whether the person making the elixir and the person taking it are one and the same, or whether there is a patient, client, or

[24] The *Rasārṇava* mentions medical cures only in passing. *Rasārṇava* 18.56 refers to a mercury-mica formulation as an agent for overcoming disease in general, strengthening the body, and increasing semen production. It gives much more detail on diseases arising through the improper use of mercurials (verses 136–140). Similarly, in the *Rasaratnākara*'s *Rasāyanakhaṇḍa*, disease is also hardly mentioned at all. Instead, the general focus is on rejuvenation and extreme longevity; creating a superhuman, invincible body that is extremely hard (*dārḍhya*) or like a diamond (*vajrakāya*); and attaining divinity or at least a comparable condition. In a very few instances, formulations are attributed with eradicating disease in general, and in three cases with curing specific disorders. See *Rasaratnakara Rasāyanakhaṇḍa* 2.121–127 for a formulation against "all diseases, ageing and death" (*sarvarogajarāmṛtyu*); 3.197–220, especially verses 215–216 for a mercury formulation that cures serious skin disorders (*kuṣṭha*), paralysis, wasted limbs, and generally all diseases; or 4.90–91 for a decoction that acts as a vermifuge and eradicates "royal consumption" and unspecified other diseases (*rājayakṣmādiroga*).

customer who is receiving treatment, or at least the elixir from an alchemist. Verse 3 speaks of the body being made to sweat by a skilled practitioner/alchemist (*vārtikendra*).[25] The indirect construction of this sentence allows for both the possibility that there is a practitioner administering treatment to another person and the possibility that it is his own body that the skilled practitioner is making sweat. In all other sentences, all actions seem to be directly undertaken by the person taking the cleansing formulations and elixirs. Using the optative mode, the text instructs that a person should undertake various activities: "One/he should eat . . .," "one/he should use . . .," and so on.

This person is variously referred to as "one undertaking the elixir regimen" (*rasāyanin*, verses 27 and 62), or as "a man" (*nara*, verse eight, or *puṃs*, verses 10 and 42), as an "intelligent or judicious person" (*matimat*, verses 33 and 45, *dhīmat*, verse 47, *sudhī*, verse 52), or conversely, as a "fool" (*mūḍha*, verse 58) in cases where the person is envisioned as not following the rules. While these appellations do not indicate whether this hypothetical person is an alchemist, we can at least state with some confidence that the *Rasahṛdayatantra*'s elixir regimen was intended for males.

In contrast to ayurvedic vitalization therapy, in which a doctor would administer treatments to a patient over time, the alchemical *rasāyana* seems to have been conceived as a course of actions undertaken by the practitioner/consumer himself. This is why the translation of "elixir regimen" is more suitable than "elixir therapy" or "elixir treatment": In alchemy, *rasāyana* is a practice.

Notes on the Text and Translation

The translation is based on the text of the 1927 edition by Yādavaśarma Trivikrama Ācārya. This edition contains the commentary *Mugdhāvabodhinī*, composed by Cathurbhujamiśra.[26] The very similar 1989 edition by Daulatarāma Rasaśāstri can be accessed online as an electronic text on the *Digital Corpus of Sanskrit*. The more recent edition and

[25] This seems to be the only instance in which a *vārttikendra*, a "skilled practitioner" or "best among practitioners," is mentioned in the *Rasahṛdayatantra*. A keyword search in the Digital Corpus of Sanskrit (www.sanskrit-linguistics.org/dcs/) shows that the term *vārttika*, "practitioner" or, given the context, "alchemist," occurs often in other alchemical texts, albeit not in the context of the final elixir intake.

[26] The date of the commentary is uncertain. Meulenbeld 2000: 622 argues that it cannot be earlier than the sixteenth century.

partial translation by Subbarayappa presents a different version of the text with many added verses and a separation of chapter 19 into three separate chapters. The additional verses and different readings have not been taken into account here.

Chapter 19 contains a few alchemical technical terms, such as *vedha*, *sāraṇa*, or *kṣetrīkaraṇa*. I have added the terms in brackets next to their translation in the text to mark them as specialist terms that are not fully conveyed by their translation. There is no standardized way of translating the alchemical terminology into English, and I have diverged a little from the usages introduced by White in *The Alchemical Body* (1996: 265–269).

The text also incorporates terms for various organic and inorganic substances. The plant materials are mostly well-known substances, but there are a few plants that I could not identify with any certainty. There is also some uncertainty in regard to the mineral substances. Notably, Hellwig's (2009) dictionary of Indian alchemy often only gives a vague categorization for substances used, such as "a kind of iron" for *kāntaloha* (which I have rendered as "magnetic iron") or for *tīkṣṇa* (rendered as "wrought iron" here), rightly pointing out the uncertainty of their proper identification. For the sake of readability, I have nevertheless chosen to provide translations of metallic and non-metallic minerals in consultation with a practitioner of Rasashastra.[27] In any case, please note that this translation is not meant to provide a do-it-yourself guide to alchemical practice: Don't try this at home!

[27] I would like to thank Dr Priyanka Chorge for providing me with information on the identification of substances in modern Rasashastra practice.

Rasahṛdayatantra, Chapter 19

1. Thus, having explained the method for transmutation (*vedha*) in the context of the rule of the king of essences, I will now speak of the mentioned formulae in the context of the elixir regimen (*rasāyana*).
2. One should drink clarified butter laced with rock salt early every morning for three days. Following this, one should use a decoction made from screwpine stilt root for three days.
3. The body should be made to sweat according to rule by a skilled practitioner. Having prepared a decoction of black hellebore, one should then undergo a thorough cleansing.
4. And following the cleansing, after all of the phlegm has been purged completely, one should partake of wholesome meals prepared from barley together with clarified butter for three days.

5–6. Further, I will explain the use of fluids for the benefit of the entire world. After having drunk hot water with ground chebulic myrobalan, rock salt, emblic myrobalan, black pepper, sweet flag, jaggery, and false black pepper in the first part of the morning, one should take turmeric and also the powder of dried ginger and of long pepper for three days.

7. Through this, one's body is cleansed, and one becomes strong and free from pathological combinations of humors. After drinking milk[28] with barley, one becomes pure.
8. Should a man undergo an elixir regimen without having prepared the body, his elixir (*rasa*) will not succeed. The elixir will cause damage to all parts of the body.
9. Thus, one who is cleansed, who has become strong through boiled rice, barley grains, and mung bean soup, and whose body has been prepared, should undertake the elixir regimen according to rule.
10. Having mixed together deodar cedar oil, clarified butter, honey, the juice of the emblic myrobalan, and milk, one should drink it and the abdomen becomes completely cleansed. The person becomes pure within.

[28] The commentary glosses *payasā* with *uṣṇodakena*, "with hot water," so perhaps hot barley water is meant instead of milk with barley.

11. In the course of a month, beauty and mental vigor arise, in two, it allays all diseases, in three months, moreover, through itself, the body becomes immortal and endowed with great vital power.
12. Drinking equal amounts of deodar cedar oil and clarified butter mixed with the juice of the three myrobalans for seven days leads to the cure of eye disorders.
13. Drinking deodar cedar oil with clarified butter, and consuming rice water and milk when the food has been digested, eliminates colds and so on along with skin diseases.
14. Deodar juice with clarified butter eliminates diseases caused by the humor bile (*pitta*), combined with oil diseases arisen from wind, with jaggery or with honey it eliminates phlegm (*kapha*)-born diseases.
15. Avoiding sour gruel and greens, one should consume rice water with milk, in a quantity of 96 grams, 192 grams, or 288 grams (two, four, or six *pala*) respectively as the lowest, medium, or highest dosage.

16–17. Following that, one should eat pure mercury that has been cleansed through distillation. Mercury that has been subjected to steaming, thickening, upward and downward distillation, strengthening, and restraining together with mica according to rule, until it becomes firm; or mercury that has been subjected [to these processes] together with pyrites.

18. The thus well-purified (*āroṭa*) mercury is first used in the preparation of the body (*kṣetrīkaraṇa*), or otherwise after it has been made into an ash, or after it has been bound through mixture with a paste.
19. One should apply this mercury for the purpose of preparing the body after it has been mixed with honey, shilajit, iron powder, chebulic and belleric myrobalans, false black pepper, ghee, and honey.
20. Thus, having eaten the mercury paste mixed together with mica, loadstone, honey, clarified butter, and so on, one attains immortality. This is the best preparation of the body.
21. Now, one should use either dark or yellow mica whose "rock-defect" has been removed after it has been repeatedly brought to a glow and sprinkle with the milky sap of camphorweed (*surabhī*)[29] for an elixir regimen.

[29] The identification of *surabhī* with camphorweed is uncertain, as the term may denote other plant materials as well, while *kṣīra* may refer to the plant's milky sap, but could also just mean milk, so that the translation would read "sprinkle with *surabhī* and milk."

22. One should eat 48 grams (one *pala*) of mica whose "shining-defect" has been removed together with false pepper, the three myrobalans, clarified butter, and honey every day, drinking milk, according to rule.
23. One should eat mica mixed with the three pungent substances, false pepper, the three myrobalans, honey and shilajit, and milk rice. A person will live a hundred years.
24. Other such preparations of sheets of mica that are mixed with sour gruel have been described many times. However, they should be avoided as they are not fit for purpose.
25. [Even in the case of] preparations of mica sheets described according to rule for the purpose of an elixir regimen, it is advised that decrepitude and disease[30] occur because of the unknown qualities of substances.
26. Mica brings one's digestive fire to cessation, as if it had never arisen. Without digestive fire, however, one is overwhelmed by a host of different illnesses.
27. The eating of mica is here said to be the very best for those undertaking an elixir regimen. Three kinds of mica are not fit for consumption: glass-like mica, contaminated mica, and "sheet-dust" mica.[31]

28–29. To begin with, mica and magnetic iron (*kāntaloha*) are rubbed with the juice of the three myrobalans. One should prepare it like collyrium covered in a cloth in the sunshine. Having made it smooth in this way, one should prepare an equal amount of various kinds of magnetic iron powder and iron and mica with false daisy many times.

30. After this nectar has been mixed with clarified butter prepared with the three myrobalans and with honey, it is placed enclosed inside two bowls under grains for a month and should be worshiped with mantras.

[30] The commentary reads *vyādhi*—disease for *mṛtyu*. *Jarā* is glossed with *pālityam*—greyness (presumably of hair), and could perhaps be translated as "symptoms of ageing."
[31] See also Hellwig 2009: 35, who notes that the commentary explains *iha* as referring to the process of *kṣetrīkaraṇa*.

31. One should retrieve it after a month, and, having ascertained its strength, one should use it, or mica without magnetic iron, or again, magnetic iron without mica.
32. It removes obesity, eye diseases like cataracts (*paṭala*), decay of sight (*kāca*)[32] and partial blindness (*timira*), swellings (*arbuda*), pains in the ears and nose, piles, anal fistula, urinary and spleen disorders and so on, and greyness of age.
33. An intelligent person who does this should predominantly consume cow milk whey, desert mung bean, clarified butter, milk, and boiled rice, observing chastity.
34. Once it has digested a quarter of the mica essence, the magnetic iron (*kānta*), and an equivalent amount of wrought iron (*tīkṣṇa*), the well-purified mercury (*āroṭa*) is applied as the best preparation of the body.
35. Mica essence, magnetic iron, and mercury, and calcined gold together with the juice of shatavari, consumed with clarified butter and honey, ward off death[33] and ageing for a year.
36. Having first made one of these remedies for the purpose of preparing the body, one should use it for a year or half a year for the attainment of ultimate bliss.

37–38. Mercury that has been killed through digesting mica, copper sulfate, iron pyrites, zinc ore, cinnabar, chalcopyrites, diamond, shilajit, and fluorite, magnetic iron, and wrought iron, and with gold, silver, brass, and copper, amalgamated or separately in groups of two, three, or four as available is recommended for an elixir regimen.

39. Mercury that is alloyed with poison, lead, or tin causes skin-diseases and so on when it is eaten. However, when mercury that is alloyed with secondary minerals[34] is eaten, all limbs shake.

40–41. [The dosage] for mercury that transmutes a hundred (times its weight) is an amount of an eighth of a gram (*guñja*) up to a

[32] *Kāca* denotes a group of eye diseases that present with loss or decay of sight without any perceptible external change in the eye.
[33] The commentary adds *vyādhi*, i.e., disease.
[34] The term *uparasa* denotes groups of mostly non-metallic inorganic substances whose common attribute is the ways in which they are processed for use and applied in elixirs. See Hellwig 2014: 466. The lists of substances considered an *uparasa* differs between the alchemical works. See Hellwig 2009: 410.

gram (*māṣa*) of mercury that has gone through the processes of "swallowing" and "digesting" mica essence, magnetic iron, iron pyrites, copper, and wrought iron; and an eighth of a gram (one *guñja*) for mercury that transmutes 1,001 (times its weight); a sixteenth of a gram (half a *guñja*) for mercury that transforms 100,000 (times its weight); and the weight of a mustard seed for mercury that transforms 10,000,000 (times its weight).[35]

42. A man who is intent on the recitation of prayers, fire oblations, and the worship of gods should carry mercury alloyed with gold, magnetic iron, and various pills.
43. [One should eat] food prepared from rice flour and clarified butter, together with mung beans and meat broth. And barley- and wheat-based foods, cow's milk, and especially whey.
44. For fluids, one should drink water and non-alkaline beverages and such sweet drinks as are recommended, laced with the four aromatic drugs and camphor, fragrant and pleasing to the mouth.[36]
45. When taking mercury, alcohol, sour gruel, sesame oil, or buttermilk are unsuitable. An intelligent person should not oil their body with pungent oil during an elixir regimen.
46. During an elixir regimen, unripe, non-sweet foods, hot milk, and spoiled meat should not be eaten. It is also taught in this context that fruit and roots that are not fresh should not be eaten.
47. "Moreover, one should not fast, but one should not eat during the night."
    One should stay away from what is forbidden. This is how an intelligent person should act during the intake of the king of essences.
48. And, having left behind thoughts and anger, one should bathe with tepid water. And one should not drive away the spirits, fever, demons and ghosts, and the divine mothers.

[35] A *guñja* is the seed of the rosary pea or jequirity bean (Abrus precatorius L.). The seeds of this plant are very consistent in weight: ca. an eighth of a gram. It is tempting to translate *guñjamātra* as "rosary pea-sized amount." However, it is not the size of the seed but its weight that is relevant here. *Māṣa* can denote a mung bean, but as a measurement of weight, it is about eight times the weight of one rosary pea seed (*guñja*), i.e., one gram.

[36] The commentary identifies the four aromatic drugs as cardamom (*elā*), cinnamon (*tvac*), nagakeshara (*nāgakesara*, the flower of the ironwood tree), and Indian bay leaf (*patra*).

49. One becomes absorbed in the highest being with a stilled mind, having achieved equanimity. Controlling one's breathing, having mastered the three goals, one becomes completely content in the bliss of mercury.
50. One who dismisses the rules of the discipline and proceeds according to his own wishes becomes confused when he takes mercury.
Because of his precarious behavior, he does not digest (the mercury).
51. If [the mercury] is not digested at this time, sleepiness, fever, faintness, burning sensations, some sharp pain in the abdomen, apathy, want of appetite, and the destruction of the body occur.
52. Having realized that it has not been digested, an intelligent person should make a medicinal preparation to cover it and give up the elixir regimen for three days.
53. If non-digestion of mercury occurs, one should drink the juice of the root of the spiny gourd and then a decoction with rock salt for three days, or with sochal salt or with cow's urine.
54. Then, having crushed sweet lime, one drinks the juice, dry ginger, and salt in the morning, boiled down together with cow's urine. It completely guards against non-digestion of mercury.
55. If, somehow or other, out of ignorance, mercury contaminated with lead was consumed, one should drink cow's urine and bitter gourd in order to remove it.
56. Wild indigo, bitter sponge gourd, pointed gourd, ivy gourd, and black nightshade: one should resort to any of the mentioned after they have been cooked when indigestion occurs.
57. The combination of mercury with substances that are too sour, salty, or bitter causes ageing. Together with overly sweet substances, and constantly consumed, the digestive fire disappears.
58. The digestive fire of a foolish person, however, who continuously does all of this, dies. Then mercury does not penetrate and disease arises.
59. However, through mercury that has been roasted at great heat and has become capable of transforming one hundred, a thousand, or one hundred thousand (times its weight), one succeeds with this procedure through effort and through the application of the mercury procedure.
60. One should eat mercury that has the transformative power of one to a hundred, a thousand, one hundred thousand, ten million, or one hundred million or more, ground with sulfur: it will give special powers.

61. Then the mercury penetrates and one will produce children radiant as divine children, and one becomes invariably loved by women, and free from wrinkles and grey hair.
62. The intelligence, strength, and power of the one undergoing the elixir regimen grows, together with his life span. The divine intelligence and divine qualities of one who has attained these grow further.
63. One who has become fully perfected through mercury, who has left behind troubles, ageing, and death and is possessed of good qualities, continually roams all the worlds through moving in the air.
64. He will also become a giver and a creator here in the triad of worlds, like the Lotus-born; one who maintains, like Vishnu, and a destroyer, like Rudra.
65. Magnetic iron, mica essence, gold and silver, and calotropis in equal shares, amalgamated with an equal amount of mercury, incinerated and made into a ball, become "lame."[37]
66. Coated externally with mercury and placed in the mouth, it protects from all weapons and destroys bodily disease. This pill is called "the immortal beauty" (*amarasundarī*).
67. The aforementioned mercury transforms more than 100,000 (times its weight). When it is bound through the application of the "potentiation" procedure (*sāraṇa*) and placed in an aperture (of an open vessel[38]), it can digest a gemstone.
68. When it is combined with equal amounts of lead and gold, magnetic iron and the essences of pyrites together with the essence of mica, the pill is called "the reviver of the dead."[39]
69. Set in gold, in a nosegay, in a tiara or a necklace, or in the ears, it combats death, fear, grief, disease, poisons, [injury by] weapons, ageing, and constant unhappiness.
70. One on whose body the pill named "reviver of the dead" is placed will be most honored by asuras, yakshas, and kinnaras and by perfected beings and eminent yogis.[40]
71. After the pill has been soaked in water for forty-eight minutes, taken out, and placed in the mouth, it will raise the dead.[41]

[37] See Hellwig 2009: 200–201 on the preparation of mercury into "lame" mercury (*khoṭa*).

[38] The commentary explains "*mukhāsthe*" with "*prakāśamukhayantra*"—an open apparatus.

[39] Gold and magnetic iron are typically not associated with having an essence (*sattva*) or subjected to the process of extracting an essence (*sattvapātana*), while pyrites are. See Hellwig 2014.

[40] *Asura*, *yakṣa*, and *kinnara* are semi-divine beings.

[41] Equals two *ghaṭika*., A *ghaṭika* is a time unit that corresponds to about twenty-four minutes.

72. A man following a salutary regimen who drinks just that water attains a divine and very firm body and becomes one who has left behind death and ageing.

73. Iron, essence of mica, and copper, gold, and silver properly alloyed and digested in equal amounts become the best of "seeds."[42] Combined with diamond, it is the "adamantine pill."

74. When placed in the mouth, it confers strength equal to nine elephants. Moreover, the body becomes impenetrable and free from death, ageing, and disease.

75–76. When one partakes of mercury that can transmute through its vapor,[43] has digested the five great gems [i.e., diamond, pearl, ruby, sapphire, and emerald] and has been "nourished" with a "seed" combined with pyrites and iron and mica essence, known as the "moving in the sky" pill, merely by placing it in the mouth, one becomes highly revered by gods, demons, and perfected beings, beginning with Indra.

77. The doctrine of mercury with its infinite qualities was arranged into the divisions of decoctions, pills, and pastes by the most eminent perfected beings. Once it has been understood by them, men prevail.

78. He whose greatness of qualities was generated through his birth in the Haihaya family of the Moon-lineage, that Honored Madana, the leader of the Kirāta tribe and scholar of mercury prevails.

79. His knowledge of alchemy transmitted itself, supplying all that is auspicious. It is the cause of the highest bliss, originally the bliss of the highest being.

80. For that reason, having obtained the high esteem of the king of the Kirātas, the monk Govinda, deeply engaging with alchemy (*rasakarman*), wrote this work called "the Heart of Mercury."

[42] A "seed" (*bīja*) is a preparation used for advanced alchemical procedures. Fed to mercury that has been processed in various ways beforehand, the seed gives the mercury the power to transmute base metals into noble ones. See White 1996: 6. According to Hellwig 2009: 79, the term *bījavara* only occurs in the *Rasahṛdayatantra*. The commentary *Mugdhāvabodhinī* on RHT 5.38 notes that *bījavara* specifically refers to a "gold seed."

[43] The commentary glosses *dhūmāvalokite* with *dhūmavedhini*. On *dhūmavedha*, i.e., transforming base metals into gold or silver through application of mercury vapor, see Hellwig 2009: 107–108.

Bibliography

Ācārya, Yādavaśarman Trivikrama. 1913. *Śrīnityanāthasiddhaviracito Rasaratnākarāntargataś caturtho Rasāyanakhaṇḍaḥ . . . saṃśodhitaḥ prakāśitaś ca = Rasayanakhanda, Fourth Part of Rasaratnâkara by Shri Nityanâthasiddha edited by Jadavjî Tricumjî Âchârya.* Āyurvedīya Granthamālā 10. Bombay: Nirnayasagar Press.

Ācārya, Yādavaśarman Trivikrama. 1927. *Rasahṛdayatantram. Śrīmadgovindabhagavatpādav iracitaṃ. Caturbhujamiśraviracita-Mugdhāvabodhinīvyākhyāsamullasitaṃ. Kāḷe ityupāhv agurunāthātmajatryambakena tathā Ācāryopāhvena Trivikramātmajena Yādavaśarmaṇā saṃśodhitam.* Lahore: Motilal Banarsidass/The Bombay Sanskrit Press.

Ācārya, Yādavaśarman Trivikrama, ed. 1981. *Carakasaṃhitā, śrīcakrapāṇidattaviracitayā āyu rvedadipikāvyākhyayā saṃvalitā.* 4th ed. New Delhi: Munshiram Manoharlal.

Ānandakanda. See Radhakrishna 1952.

Aṣṭāṅgahṛdayasaṃhitā. See Srikantha Murthy 1999–2000.

Aṣṭāṅgasaṃgraha. See Srikantha Murthy 1995–1997.

Cakradatta. See Sharma 2002.

Carakasaṃhitā. See Ācārya 1981.

Hellwig, Oliver. 1999–. *DCS: Digital Corpus of Sanskrit.* http://www.sanskrit-linguistics.org/dcs/.

Hellwig, Oliver. 2008. "*Rasāyana* und die āyurvedische Krankheitskunde." *Traditional South Asian Medicine* 8: 32–64.

Hellwig, Oliver. 2009. *Wörterbuch der mittelalterlichen indischen Alchemie.* Groningen: Barkhuis & University of Groningen, University Library.

Hellwig, Oliver. 2014. "Materialgruppen in der indischen Alchemie." *Zeitschrift der Deutschen Morgenländischen Gesellschaft* 164 (2): 451–468.

Maas, Philipp. A. 2017. "Rasāyana in Classical Yoga and Āyurveda." *History of Science in South Asia* 5 (2): 66–84, https://doi.org/10.1873/hssa.v5i2.32.

Mazars, Guy. 2006. *A Concise Introduction to Indian Medicine.* Translated by T. K. Gopalan. Delhi: Motilal Banarsidass.

Meulenbeld, Gerrit Jan. 2000. *A History of Indian Medical Literature.* Volume 2A. Groningen: E. Forsten.

Nadkarni, A. K. 1954. *Dr. K. M. Nadkarni's Indian Materia Medica, with Ayurvedic, Unani-tibbi, Siddha, Allopathic, Homeopathic, Naturopathic & Home Remedies, Appendices and Notes.* 2 vols. Bombay: Popular Prakashan.

Radhakrishna, S. V., ed. 1952. *Anandakandam, Edited with Translation in Tamil, and Introduction in Tamil and Sanskrit.* Volume 15. TMSSM Series. Thanjavur: Tanjore Maharaja Serfoji's Saraswati Mahal Library.

Rasahṛdayatantra. See Ācārya 1927; Subbarayappa 1997; and Rasaśāstri 1989.

Rasaratnākara Rasāyanakhaṇḍa. See Ācārya 1913.

Rasaratnasamuccaya. See Vidyābhūṣana 1927.

Rasārṇava. See Ray 2007.

Rasaśāstri, Daulatarāma. 1989. *Rasahṛdayatantram. Śrīmadgovindabhagavatpādaviracitaṃ. Caturbhujamiśraracita "Mugdhāvabodhinī" Saṃskṛta ṭīkā tathā Hindī vyākhyā sahitam, vyakhyākṛtṛ Daulatarāma Rasaśāstrī.* Vārāṇasī: Caukhambhā Oriyanṭāliyā

Ray, P. C., and H. C. Kaviratna. 2007. *Rasārṇava. Edited by P. C. Ray, H. C. Kaviratna. Re-Edited by S. Jain.* Delhi: Oriental Book Centre.

Roşu, Arion. 1975. "Considérations sur une technique du Rasāyana āyurvédique." *Indo-Iranian Journal* 17 (1): 1–29. htps://doi.org/10.1163/000000075790079222.

Sharma, Priya Vrat. 2002. *Cakradatta (Sanskrit Text with English Translation): A Treatise on Principles and Practicesof Ayurvedic Medicine.* Varanasi: Chaukhamba Publishers.

Siddhayoga. See Tewari 2006.

Srikantha Murthy, K. R. 1995–1997. *Aṣṭāṅga Saṃgraha of Vāgbhaṭa, Text, English translation, Notes, Indices etc., translated by K.R. Srikantha Murthy*. 3 vols. Varanasi: Chaukhamba Orientalia.

Srikantha Murthy, K. R. 1999–2000. *Vāgbhaṭa's Aṣṭāṅga Hṛdayam, Text, English translation, Notes, Appendix and Indices, translated by K.R. Srikantha Murthy*. 3 vols. Varanasi: Krishnadas Academy.

Subbarayappa, B. V., B. S. Ramakrishna Rao, M. Mahadeva Sastry, and S. R. N. Murthy. 1997. "Rasahṛdaya Tantram of Govinda Bhagavatpāda." *Indian Journal of History of Science* 32 (1): 1–68, 32 (2): 69–148.

Tewari, Premvati, ed. 2006. *The First Treatise of Āyurveda on Treatment Vṛndamādhava or Siddha Yoga*. Varanasi: Chaukhambha Visvabharati.

Vidyābhūṣaṇa, Āśubodha, and Nityabodha Vidyāratna, eds. 1927. *Saṭīkaḥ Rasaratnasamuccayaḥ: (prācīnarasagranthaḥ)*. Kalikātā: The editors.

White, David Gordon. 1996. *The Alchemical Body: Siddha Traditions in Medieval India*. Chicago: University of Chicago Press.

Wujastyk, Dagmar. 2013. "Perfect Medicine. Mercury in Sanskrit Medical Literature." *Asian Medicine* 8: 15–40.

Wujastyk, Dagmar. 2017. "Acts of Improvement." *History of Science in South Asia* 5 (2): 1–36. https://doi.org/10.18732/hssa.v5i2.26.

Wujastyk, Dominik. 2003. *The Roots of Ayurveda: Selections from Sanskrit Medical Writings*. 3rd ed. Penguin Classics. London and New York: Penguin Group.

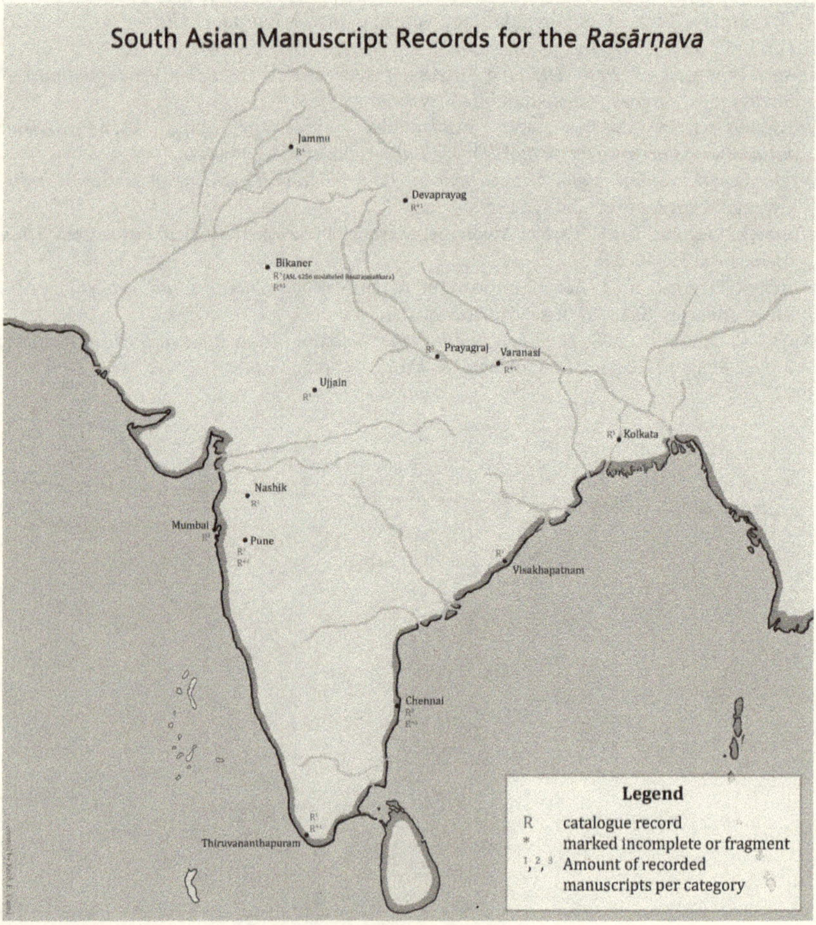

Map 2.1 South Asian manuscript records for the *Rasārṇava*.

2

The Practice of Alchemy and Its Practitioners

Chapters 1 and 2 of the *Rasārṇava*

David Gordon White

The eleventh-century *Rasārṇava* ("Ocean of Mercury") is a foundational guide to Hindu alchemy wrapped inside a Hindu tantric scripture. Its tantric tone is set in the work's opening verses, in which the quintessential tantric god Bhairava is identified as the preceptor of the Kaula. Derived from the noun *kula* ("clan," "lineage," "swarm"), the Kaula was a current of Hindu Tantra that, in addition to incorporating the use of transformative sexual fluids into its practice, enshrined one or another form of Bhairava at the heart of its otherwise mainly feminine pantheon, and placed significant emphasis on the attainment of supernatural powers (*siddhi*s) in addition to the more conventional goals of identity with or eternal proximity to Śiva, and liberation at the end of life.[1] While much of the *Rasārṇava*'s discussion of initiation and related matters is modeled after tantric scriptures compiled in Kashmir in the seventh to ninth centuries, the work's rare geographical references indicate that it was compiled in the eastern part of the modern-day Indian state of Maharashtra.[2] In fact, a number of India's principal alchemical works hail from the same Deccan plateau region, and as a recent study has demonstrated on the basis of extensive textual, iconographic, and epigraphical data, the religious mainstream of this wide region was, down through the end of the twelfth century, Kaula.[3]

The *Rasārṇava* is, like most "revealed" tantric texts (called Tantras), cast in the form of a dialogue in which the Goddess (here called Pārvatī, Devī, or

[1] Identity with Rasabhairava is implied in 2.85b, and as Brunner (1975: 418–420) has shown, the *sādhaka*'s immediate goal of supernatural powers in this life does not exclude identity with some form of Śiva, realized through initiation.
[2] White 1996: 113–114.
[3] Schwartz 2023.

David Gordon White, *The Practice of Alchemy and Its Practitioners* In: *Indian Alchemy*. Edited by: Dagmar Wujastyk, Oxford University Press. © Oxford University Press 2025. DOI: 10.1093/9780197805831.003.0003

Bhairavī) asks the questions and Śiva (also called Bhairava here) answers them. Verses 1–6 set the scene and introduce one of the main themes of the work's opening chapter: the quest for bodily liberation through alchemical practice. Bhairava begins his answer by stressing the concrete nature of the alchemical path: in contradistinction to the teachings of more conventional Hinduism (the "Six Schools" of Indian philosophy), which maintain that liberation occurs after the shedding of one's mortal body, here it is this body of flesh and blood that is transformed and rendered immortal and unageing (1.8–9, 12–17, 21).

The author reserves his greatest rhetorical ammunition for the practices of other tantric sects that, he clearly feels, are on the wrong track. These he condemns for their antinomian practices grouped under the common heading of the five *makāras*, the five "M-words," comprising flesh (*māṃsa*), fish (*matsya*), alcohol (*madhu*), sexual intercourse (*mudrā*), and sexual fluids (*maithuna*) as means to realizing tantric goals. (1.10–11, 24, 29–30). The author appears to contradict himself regarding his anti-*makāra* position in verse 25, in which Bhairava praises eating "beef" and drinking "liquor." This, however, is a cryptic reference to a technique known as *khecarī mudrā* ("the seal of moving in the ether"), a component of the body of breathing techniques, postures, locks, and seals that comprised the emergent tradition of *haṭha yoga*, the "yoga of violent exertion."[4]

Perfected mercury serves as a catalyst for the internal transformation of the gross mortal body into one that is perfected and immortal (1.18–22). As verse 21 indicates, this is a natural process, comparable to cognition. Alchemical and yogic practice alone are not sufficient, however: without the initiation, divine grace (issuing in *śaktipāta*, the "descent of the Goddess"), and instruction by a spiritual preceptor (*guru*), the alchemist will fail in his quest for power, pleasure, and immortality (1.23, 54–57).

Much of the *Rasārṇava*'s first chapter is devoted to the greatness of mercury itself. It is Śiva's seed (1.36) or the mineral product of the sexual union of Śiva and the Goddess (1.34). Itself possessed of wondrous powers, mercury transmits said powers to the alchemist who ingests it (1.19, 27–28, 31, 53, 57). All of the principal terms for mercury are found in this opening chapter: *rasa* (about which more below), *rasendra* ("Champion of Rasas"), *sūta(ka)* ("that

[4] In the practice of *khecarī mudrā*, the practitioner pushes his tongue up into his sinuses ("eating beef") in order to imbibe the fluid that his yogic practice has made to flow ("drinking liquor"): *Haṭhayogapradīpikā* 3.48–49, with the commentary of Brahmānanda, in Iyengar 1972: 89–90 (Sanskrit); 45–46 (English).

which was born [from Śiva's semen]"), and *pārada* (for which a problematic etymology is offered in verse 35). Beyond being equated with Śiva's semen, mercury is further identified with the god's phallic emblem, the *liṅgam* from which it arose in the beginning. For this reason, the construction and worship of mercurial phalluses (*rasaliṅgams*) are central to alchemical practice, just as *liṅgam*-worship is central to devotional cult of Śiva, for which such *liṅgams* as Kedārnāth in the Himalayas are important pilgrimage sites (1.37–43).[5] Abuse of or lack of faith in mercury is, in this context, a sacrilege, the punishments for which are detailed in verses 45 through 52.

The importance of the alchemical preceptor (*guru*), underscored in verses 54 through 59, is introduced by a reference to the rays (*marīcayaḥ*) that serve as the luminous conduits for the transmission of his teachings. Attested from the time of the Hindu epics, rays are widely evoked as transfer media for exchanges among humans, gods, and the physical world.[6] The chapter concludes with a reference to a tantric ritual technique called *nyāsa*, in which mantras of the god and goddess of alchemy are deposited upon the practitioner's body, in preparation for self-divinization. The final verse (1.60) opens the way to the remaining seventeen chapters of the *Rasārnava*, devoted in the main to alchemical "recipes" and descriptions of the wondrous powers to be gained through the use of perfected mercury.

It is, however, the *Rasārṇava*'s second and third chapters, on the topics of initiation (*dīkṣā*) and the depositing of mantras (*mantra-nyāsa*), that enshrine the greatest concentration of tantric material in the entire work. The tone is set in the opening verses of chapter 2, when Bhairava describes the ideal alchemical guru as a person devoted to the "path of the tantric clan" (*kula-marga*) (2.2), and the ideal initiand as "devoted to the veneration of . . . the circle of Yoginīs and the clan [deities]" and a "cultivated connoisseur of Tantra" (2.8). In the course of this chapter, a host of deities from the Hindu tantric pantheon will be invoked, together with a palette of tantric practices: the fashioning of ritual diagrams (*maṇḍalas*); the use of spells (*mantras*, *vidyās*), seals (*mudrās*), menstrual blood, and other power substances; tantric worship; meditation; protective rites; and so forth.

In its colophon, the *Rasārṇava*'s second chapter carries the title of the "Rules for Initiation" (*dīkṣāvidhāna*). This chapter's terminology, program of initiation, and descriptions of the supreme divinities at the heart of its

[5] White 1996: 245–246.
[6] White 2009a: 122–170.

worship *maṇḍala*—their iconography, mantras, and so forth—provide us with several clues concerning the *Rasārṇava*'s place in the medieval Hindu tantric universe. Even if its initiation program is a simplified version of those found in them, its anonymous author is clearly taking his inspiration from such early Kashmirian tantric scriptures as the *Siddhayogeśvarīmata*, *Svacchanda Tantra*, and *Netra Tantra*.[7] Its account of initiation is patterned after these works in many respects, including its description of a suitable initiation site (2.37–47); the construction of an initiation *maṇḍala* for the worship of the deities of the tantric pantheon (2.48–73a); tribute offerings for the pacification of noxious spirit beings (2.73b–76b); the anointment of the initiand with mantra-incanted water poured from a consecration vase (2.78–81); the transfer of the initiand's subtle body and breath onto a cotton cord (*pāśasūtra*) tied to the initiand's topknot during initiation (2.84b);[8] the nocturnal celebration of a Kaula circle-of-worshipers rite (*cakrapūja*) involving human virgins, yoginīs, yogis, and initiands (*sādhaka*) (2.82–83); and the depositing of mantras on the body (2.93, 97–98). A certain number of standard features of tantric initiation, such as the casting of a flower onto the *maṇḍala* as a means to determining the initiand's new tantric identity, are, however, absent from the *Rasārṇava*'s discussion. At the same time, the chapter contains a number of observations that are tangential, if not irrelevant, to initiation: discussions of the body types of various alchemical assistants, both male and female (2.9–11, 17–26), mirror divination (2.92–103), and advice to the alchemist concerning his general behavior and ways of treating perfected mercury (2.122–132). While much of the style of the second chapter follows tantric scriptural norms—ritual instructions delivered in a kind of shorthand that assumes prior knowledge on the part of the reader or hearer—certain of its passages aspire to a higher level of expression. In its description of the ideal female laboratory assistant (2.20–25), its portrayal of the wonderland surrounding the alchemical laboratory (2.41–43), and its allegorical enumeration of various worship paraphernalia (2.86–87), we see its author attempting to adopt the embellished style of Sanskrit

[7] Initiation (*dīkṣā*) is the principal topic of chapters 6–8 of the *Siddhayogeśvarīmata* (in Törszök 1999: 121–142), chapters 4–5 of the *Svacchanda Tantra* (in Shastri 1923: 1–332; Shastri 1926a: 1–99), and chapter 4 of the *Netra Tantra* (in Shastri 1926b: 106–115). The *Rasārṇava*'s second chapter was the object of extensive borrowing by a number of later alchemical works: the *Rasaratnākara* of Nityanātha (part 3 [Ṛddhikhaṇḍa], chapter 1), the *Rasaratnasamuccaya* of Vāgbhaṭa II (chapter 6), the *Rasendracūḍāmaṇi* of Somadeva (chapter 2), and the *Ānandakanda* of Mahābhairava (part 1, chapters 2–3).

[8] Davis 1991: 94–100.

poetical composition (*kāvya*), in one case rather awkwardly inserting a *subhāṣita* aphorism into his narrative.[9]

Here, as in virtually every South Asian ritual since the time of the Vedas, mantras are the motors of practice. No element of ritual—whether it be an implement, an offertory substance, or, most crucially, a god—can be brought into the ritual sphere, activated, or made efficacious without its corresponding mantra. By the same token, no ritual act, including initiation, can be performed without mantras, whose depositing (*nyāsa*) onto the body of the initiand and various ritual supports is implemented through the use of various seals (*mudrās*), devices (*yantras*), and so forth.

Sexual fluids are brought to the fore in the second chapter's discussion of *rasabandhakīs*, the alchemist's female "fixa-trixes" of mercury (2.17–26). Classified according to the timing of their menstrual cycles, these female types—two of whose names (*kākiṇī* and *padminī*) are derived from poetic terms for the vulva, described here in voyeuristic detail—are vital to the alchemist inasmuch as their menstrual emissions constitute an "essential fluid" (*rasa*) for the fixation of mercury. The *Rasārṇava* evokes such females in two other chapters, explaining (14.2) that the menstrual blood of a woman who consumes sulfur (the mineral equivalent of divine and human menses) for twenty-one days is a catalyst for the fixation and calcination of mercury. In its final chapter (18.165–169), the work describes the female charms of the *kāminī* ("passionate damsel") in terms similar to those employed in chapter 2 to describe the *padminī*. The *kāminī*'s sexual favors and embrace—or simple presence—render the alchemist impervious to weapons, turn his grey hair to black, remove wrinkles, and endow him with various supernatural enjoyments. Interestingly, the reverse case, in which a female (generally a superhuman Yoginī) is empowered by the mere presence of a male—although eating said male is also an option for her—is also alluded to in this chapter, when it speaks of a person who has had "seven rebirths" (2.104). Both Hindu and Buddhist Tantras know of males who, having been the sacrificial victims of the Yoginīs in six former lives, afford immediate *siddhis* to any Yoginī who would happen to kill, eat, or even touch or talk to them.[10]

If, as some have argued, Tantra was a vehicle for the empowerment of human women, in these cases their "sacred sexuality" rather appears to

[9] Verse 2.35b is a paraphrase of *sahāyāstādṛśā eva yādṛśī bhavitavyatā*, which appears as nos. 3732 and 4122 in Böhtlingk's 1865 compendium.

[10] Serbaeva 2013: 201; Serbaeva 2016: 70–71nn68 and 69; Gray 2005: 65; Gray 2007: 206–207, 219, 367–368.

have been a means for their instrumentalization. The tantric glorification of the divine feminine did not necessarily translate into the social uplift of human females, as the verses that follow the account of the "fixa-trixes" (*rasabandhakī*) make plain. Here it suggests that the sole effective way to find a woman who is both attractive and enthusiastic, and who menstruates like clockwork, is to create one. This is effected through a woman's repetition of a mantra and her employment of a *mudrā*, which together bring about her (?immediate) "birth" as a living goddess, a daughter of the tantric divinities Sadāśiva and Caṇḍikā (2.31–34). This practice, which I have not seen referred to elsewhere in the alchemical or tantric canons, is declared by the *Rasārṇava* to be a "supreme mystery." Here, one should recall that the Hindu gods do not reproduce sexually. They are themselves "not born from a womb" (*ayonija*), and their rare divine offspring tend to be procreated through spermatogenesis (Skanda-Kārttikeya), parthenogenesis (Gaṇeśa), or embryo transfer (Kṛṣṇa).

The *Rasārṇava*'s Kaula cum alchemical pantheon is embedded in its initiation *maṇḍala*,[11] constructed around a mercurial *liṅgam-yoni* placed at the center of a brick worship altar. That *liṅgam*, which is identified with Rasabhairava, the "Mercurial Bhairava," is also the support for the great god's consort, named Rasāṅkuśī ("Elephant Goad of Mercury"), first mentioned in the penultimate verse of chapter 1. Both are established in the *liṅgam-yoni* through the utterance of mantras, in this case the *rasāṅkuśī* mantra and the "root mantra of King Mercury" (2.49–53). Following its instructions for their installation, our text provides a detailed account of an eleventh-century alchemist's tantric pantheon, beginning with a group whose names are variations on those commonly found in lists of Śiva's eight *gaṇeśvaras* ("Masters of the Minions") (2.54), lists that include the names of Nandin, Mahākāla, and Bhṛṅgin. While Bhṛṅgin is doubled here into Bhṛṅgirīṭa and Bhṛṅgīka, and Sugrīva and Kumbhakarṇa are names of powerful figures from the *Rāmāyaṇa*, Mahābala ("Very Mighty") and Dṛḍhāyudha ("Having Strong Weapons") are both generic sorts of names for powerful door guardians (2.54a–55a). Moving inward, the eight directional guardians are next installed on the peripheral quadrants of the maṇḍala (2.55b–56a). Here as well, the names of several of these divinities are at variance with those found on standard lists.[12]

[11] A nearly identical *maṇḍala* is described in the *Rasendracūḍāmaṇi* (2.21–34), whose author Somadeva smooths over many of the irregularities found in our text. See Roşu 1986: 251 for a graphic representation of this *maṇḍala*.

[12] The standard names of this octad are found in 2.107–108.

Our text next describes a set of three concentric lotus flowers, whose petals are the supports for tantric deities, "transitional" female entities, and divinized alchemical processes and reagents. Gracing the petals of the outermost lotus are two sets of ten female entities (2.57–58). The first of these is comprised of names for alchemical processes and supernatural powers, feminized through the addition of feminine -ā or -ī endings. This set is paired with the ten "female messengers" (*dūtīs*), a term generally employed for tantric consorts who transmit the transformative "clan substance" (*kula-dravya*) to male initiates through their menstrual emissions. However, several tantric traditions simply identify the *dūtīs* as minor female deities in the entourage of Bhairava and his consort.[13] Inscribed inside of this flower is a second lotus blossom, whose eight petals are the supports for the eight "male" primary reagents (*mahārasa*).

The petals of the innermost flower are occupied by the eight "Masters of Wisdom" (*vidyeśvaras*), fixtures of the Śaivasiddhānta pantheon (2.59–61).[14] These ring the lotus's pericarp, inhabited by the four supreme goddesses of the Kashmirian Trika Kaula: Mālinī ("Garland [of Letters]"), Parā ("Transcendent"), Aparā ("Non-Transcendent"), and Parāparā ("Transcendent-Non-Transcendent") (2.62). Surrounded by the four Trika goddesses at the heart of the mandala is the dyad of Rasabhairava and Rasāṅkuśī, whose respective iconographies and mantras closely resemble those of Svacchandabhairava as described in the *Svacchanda Tantra*, and the Bālā Tripurasundarī of the Śrīvidyā Kaula, respectively.[15] Both are highly auspicious and relatively unterrifying in appearance (2.63–73). As for Rasāṅkuśī's name, it may be explained on the basis of Bālā Tripurasundarī's iconography, which shows her holding an elephant goad in her upper right hand;[16] the special relationship of the elephant goad to alchemical practice is explained in this chapter's verse 90.

[13] For discussion, see White 2003: 110–114. In his *Tantrāloka*, Abhinavagupta's *siddha-cakra* features four *dūtīs*, who are the consorts of the Siddhas who receive and disseminate the Kaula revelations in the four world ages (*yugas*). These four are also mentioned in the *Netra Tantra*, while the *Kubjikāmata* describes a *dūtī-cakra* comprised of eighty-one divine *dūtīs*. Other works refer to a grouping of six *dūtīs*.

[14] The eight Vidyeśvaras, "pure beings who are assigned by Śiva to reign over the entire impure domain . . . the first and most direct agents delegated to carry out Śiva's commands" (Davis 1991: 66) are: Ananta, Sūkṣma, Śivottama, Ekanetra, Ekarudra, Trinetra, Śrīkaṇṭha, and Śikhaṇḍin. These names vary slightly from one source to another.

[15] Personal communication from Alexis Sanderson, July 6, 1992. For further discussion, see White 1996: 178 and 443nn62–63.

[16] Sanderson 1988: 688 and fig. 36.8.

Following the description of its initiation *maṇḍala*, our text enters into a summary discussion of the initiation ritual proper, beginning with the offering of tribute (*bali*) to a group of malign spirit beings that would otherwise obstruct the ritual proceedings. They are so inclined because, as the original *genii locorum* of the land upon which the ritual is taking place, they consider such tribute to be their due. Furthermore, simply due to their obnoxious nature, they are always in need of "pacification," which here results in their "removal," if not their "destruction" (*nāśanam*). The "obstructions" named here are, more properly speaking, Obstructers, since they are sentient beings possessed of free will. Their various sub-types are frequently encountered in these contexts: the Trespassed (*bhūta*s), Dryads (*yakṣa*s), Guardians (*rākṣa*s), and Flesh-Eaters (*piśāca*s) (2.119–120).

Although the term is not employed, yet another *maṇḍala* is implied later in the chapter. Here, following its account of the apprentice's (*sādhaka*) initiation, our work describes the layout of the alchemist's laboratory, once again bringing the eight directional guardians into play. In this case, the gods' names and functions relate directly to the alchemical reagents and procedures assigned to their placement. So, for example, metal-killing reagents are stored in the south, the direction of Yama, the god of death; and the bellows are installed in the northwest, the direction of Vāyu, the wind god (2.107–108). Following this, the newly initiated practitioner is instructed to install his guru on a throne at the center of the laboratory. At this point, a long passage identifies the Kaula universe with the cosmic body of Rasabhairava, which is presumably identified with that of the guru himself, upon which the practitioner is instructed to meditate (2.109–117). Beginning with the names of the five faces of the Sadāśiva,[17] superimposed upon the lower body of Rasabhairava from his knees to his eyebrows, the work goes on to identify ever higher levels of being/consciousness, extending beyond the upper contours of the god's "physical" body and corresponding to the upper reaches of the metaphysical systems of the *Svacchanda* and *Netra Tantras*.[18]

The importance of mantras is brought to the fore in the context of a divination rite. Here, following verse 97, two long unnumbered passages detail

[17] The supreme godhead of the Śaivasiddhānta system, who is also the "ghost" (*preta*) upon whom the divine pair are seated in 2.64, Sadāśiva is relegated to lower metaphysical/iconographic levels in the Kaula systems through the strategy of what Sanderson (1986: 181–185) has termed "superenthronement."

[18] In these two works, discussed in Padoux 1989: 404–411, the levels above the eyebrows are identified with the subtle components of the Oṃ mantra.

the garland (*mālā*) and root (*mūla*) mantras, which actuate the depositing (*nyāsa*) of six fearsome Bhairavas onto the practitioner's hand and "members" (*aṅgas*) of the body. As is commonly the case in the Tantras, some of these members are not parts of the physical body, but rather parts of a ritually constituted tantric bodily Gestalt. These include the *kavaca* ("armor," or "encasement" in tantric parlance[19]) and *astra* ("missile," a term that in this context refers to a type of mantra). As for the garland mantra, the names of the deities invoked in it, while generic, are evocative of the Kaula "culture of the cremation ground": Great Black One, Very Powerful, Unterrible, Adamantine Hero, Anger, and Skeleton. In addition to being deposited on the corners of the hexagonal mandala described in verse 96, these are successively deposited on the two thumbs (Mahākāla), two index fingers (Mahābala), and so on, with Kaṅkāla deposited on the palm and back of the hand. Through their *nyāsas*, these Bhairavas encase the practitioner's body, protecting its every part and transforming it into a divine body.[20]

Next follows the description of the divination rite proper, which is significant inasmuch as it is one of the very rare accounts of a South Asian divinatory practice that likely originated in Hellenistic Egypt in the early centuries of the Common Era. First appearing in Indic texts in about the seventh century, the full-blown ritual brings three actors into play: an adult magician, his child assistant, and a deity called down onto a reflective surface or lamp. *Praśna* ("questioning"), the Sanskrit term for the practice, came to be feminized in Indic sources as Praśnā, the name of the spirit or minor deity made to descend (*ava-tāraṇa*) onto the surface of the divinatory apparatus.[21]

Apart from this account of the ritual descent of Praśnā, another passage unique to the *Rasārṇava* concerns the "body types" of the male specialists the alchemist will employ for what are referred to as the "treasure practice" (*nidhi-sādhana*), "pit (or portal) practice" (*vila-sādhana*), and "smelting" (*dhātuvāda*), respectively (2.9–11). These verses appear to be referring back to *Rasārṇava* 1.44's discussion of four "doctrines" (*vādas*): in ascending order, these are (1) *khaga-* ("sky-going," but perhaps a misspelling of *khaḍga-*, "sword," or *khanya-*, "excavational"), (2) *vila-* ("aperture," "portal"), (3) *mantra-* ("spell"), and (4) *rasa-* ("alchemical").[22] Three tales from the

[19] The term refers to small protective amulet cases threaded into cords worn on the body: Verardi 1994: 11 and fig. 16.

[20] The six *aṅga-mantras* and their respective "limbs" are discussed in several tantric works. See Brunner 1986: 89–132.

[21] Discussed in White 2021: 73–74.

[22] These "doctrines" and their attendant practices are discussed in Vasudeva 2012: 272–279.

Kathāsaritsāgara, an eleventh-century Kashmirian anthology, involve occult specialists (two of whom are tantric Pāśupatas) of *nidhi-prasādhana*, the practice of finding hidden treasure.[23] In one of these, *bila*, *khaḍga*, and *khaga* practice are combined: a Pāśupata aids a king in finding a portal (*bila*, i.e., *vila*) that leads him to a hidden sword (*khaḍga*) belonging to a female Asura: she lovingly offers it to the king, assuring him that it will afford him the power of flight (*khagati*). This squares with Somadeva Vasudeva's identification of the goal of *bila* practice as passing through portals into subterranean worlds inhabited by just such sexually gratifying supernatural damsels.[24] In its chapter on *pātala-siddhi*, the *Jayadrathayāmala*, a foundational tantric scripture, lists seventeen specific geographic locales in the Kashmir region featuring special portals through which these worlds may be accessed.[25]

It would appear, however, that verse 10 of our chapter is interpreting *vila-sādhana* in a different way, with the massive brute it describes being more suited to pit-mining than to finding hidden portals to enchanted underworlds, in which case *khanya* ("excavational") would be the preferred reading. *Nidhi-sādhana*, the art of finding hidden treasure, involves sleuthing techniques as well as certain innate gifts or powers. By referring to specific markings on the hands of its specialists, our chapter's verse 9 appears to be alluding to such specialized aptitudes.[26] However, one of the tales from the *Kathśaritsāgara* employs the terms *khanya-vādin* and *nidhāna-vādin* (a variant form of *nidhi-vādin*) interchangeably, an indication that the same persons who detected hidden treasures also dug them out of the ground.[27] Whereas *dhātuvāda* (literally the "doctrine of metals") often simply denotes metal-based "alchemy," it appears here to refer to the practical art of metallurgy—or, judging from the description given in verse 11, to specialists in the technique of smelting (*lohādhikaraṇam*).

These are not the sole terms that pose challenges to the translator. Several others must be mentioned here, beginning with two of the most ubiquitous nouns in this chapter: *rasa* and *sādhaka*. When given in the singular, *rasa*, the Sanskrit cognate of the English word "resin," most often refers to mercury. However, in one case, the word instead refers to the "essential fluid"

[23] *Kathāsaritsāgara* 6.8.67–78; 9.6.214–219; 10.5.36–38.

[24] Vasudeva 2012: 275.

[25] Sanderson 2004: 280–283.

[26] For additional background on *nidhi* practice in scriptural and literary sources, see Balbir 1993: 15–53.

[27] *Kathāsaritsāgara* 6.8.69,73, discussed in Balbir 1993: 19.

generated by the alchemist's female assistant on the phases of the moon (2.25b). Elsewhere in this chapter (2.121, 124–126), I have translated another term employed to denote perfected mercury (*dravyam*, literally "substance") as "the elixir." The plural *rasa*s are mineral reagents, with the *mahārasa*s (2.4b) being the primary reagents used in the alchemical Work (my translation of the Sanskrit *karman*, in both the singular and the plural).

As for *sādhaka*, this word has a number of general and specific senses, which our text shuttles between indifferently. In the first portion of this chapter, down to the conclusion of the initiation ritual proper at verse 83, I have translated *sādhaka* as "initiand" and "apprentice," both terms being appropriate designations for a person aspiring to practice alchemy in his own right. Following verse 83, however, the *sādkhaka* is more properly understood to be the now initiated "practitioner," and more specifically (in line with broader tantric categories of initiation), a ritual technician whose primary goal is the attainment of *siddhis*[28] ("supernatural powers") in this life. In his quest for such powers, the *sādhaka* frequently interacts with female figures called *yoginīs*.[29] In one case, in which these are cast as divine members of the Kaula pantheon, I have given their name in the uppercase (2.8); in another, when they are mentioned together with other human practitioners because they are themselves human, their name is lowercase, italicized (2.82).

Finally, a word on the relative absence of "alchemy" in this chapter of an alchemical text. In spite of its focus on tantric initiation, chapter 2 does mention a number of alchemical processes (*saṃskāra*s, in 2.4–6 and 109); as well as alchemical reagents, both mineral and botanical; the alchemical laboratory; and the alchemist himself. And, of course, miraculous mercury is everywhere, both as a chemical, as a *liṅgam*, and as the name of the supreme divinity of an enchanted alchemical universe.

Notes on the Text and Translation

The *Rasārṇava* has been edited twice. P. C. Ray and Hariscandra Kaviratna's non-critical 1910 edition is based on five manuscript (MSS) witnesses, from Bikaner, the Deccan, Kashmir, Madras, and Mysore. The editors give no

[28] In 2.131, I have translated *siddhi* in the singular in the general sense of "success." On the semantic range of this term in specific tantric contexts, see Brunner 1975.

[29] "Female Joiners": for an explanation based on *Netra Tantra* 20.4–21, see White 2009b: 132 and White 2024: 373–376.

further information on their manuscript sources, such that, for example, one cannot know whether their Bikaner manuscript was one of the three that I consulted in the Anup Sanskrit Library in the winter of 1992: MSS 4256, 4273, and 4274. In a few cases, referenced in footnotes, I have preferred the readings of these manuscripts over those found in Ray and Kaviratna's edited volume. Indradeo Tripathi's 1978 edition, based on that of Ray and Kaviratna, provides a rough Hindi-language translation together with Hindi commentary, notes, and a helpful ninety-page lexicon of alchemical terminology. A digital version of chapters 1 through 12 of Ray and Kaviratna's edition is online at http://gretil.sub.uni-goettingen.de/gretil/1_sanskr/6_sastra/7_ayur/rasarnau.htm (input by Oliver Hellwig). Chapters 1 through 18 are available on line on the Digital Corpus of Sanskrit at www.sanskrit-linguistics.org/dcs (input also by Oliver Hellwig).

Rasarṇava, Chapter 1

1 He in whom everything is, from whom everything will be, who is everything and everywhere, who is all-encompassing and eternal: Salutations to that universal soul!

2–3 While atop Kailāsa's delightful peak, a peak adorned with a multitude of jewels and strewn with myriad trunks and vines, in a secret spot where none of their fellow gods were present, the goddess Pārvatī, bowing her head before the blue-throated, three-eyed god of gods who was comfortably seated there, asked:

The venerable Goddess spoke:

4 O god among gods, O great god, O incinerator of Time and Love, O preceptor of the Kaula, Mahākaula, Siddhakaula, and other lineages![30]

5–6 By thy grace, all of revelation has been made known to the world, in its entirety. If I am worthy of thy favor, if I am thy beloved then tell me this: there are things that have been communicated in all of the Tantras, yet there are things that have not been brought to light. O Lord! What is this [thing called] liberation in the body (*jīvanmukti*)? Thou alone art capable of describing that.

The venerable Bhairava spoke:

7 Well said, well spoken, O fortunate Goddess! Well spoken, O daughter of the mountain! The question thou hast posed is a matter that concerns the welfare of our devotees.

8–9 Eternal youth, immortality of the body, and the attainment of an identity of nature with Śiva—that is, liberation in the body—is difficult even for the gods to attain. The liberation that occurs when one drops dead is indeed a worthless [sort of] liberation. [For in that case] a donkey would also be liberated when he drops dead.

10 If liberation were [simply] the excitation of the female genitalia, whom could donkeys not liberate? Indeed, what cow or she-goat would not be liberated, O Little Mother of My Minions?

11–13 Therefore, one should safeguard one's body with mercury and mercurial preparations. If liberation came from enjoying semen,

[30] I have emended *siddhakaulādināśana* to read *siddhakaulādiśāsana*, on the basis of these manuscript witnesses from the Anup Sanskrit Library, Bikaner: ASL MSS 4256, 4273, and 4274.

urine, and excrements, which of the races of dogs and swine would not be liberated? Liberation is indeed viewed in the six schools as occurring when one drops dead, but that [sort of] liberation is not directly perceptible, in the way that a myrobalan fruit in the hand [is perceptible]. Ineffable reality, even that shall I declare to thee, o Goddess!

14–16 In no wise can a person who is merely without sin or worshiping with mantras preserve his body. If the preservation of the body is difficult even for the gods, O great Goddess, then how much more difficult it must be for men living on the face of this earth! When righteousness has disappeared, where can there be righteousness? When righteousness has disappeared, where is practice? When practice has disappeared, where is the Work? When the Work has disappeared, where is the stratagem? When the stratagem has disappeared, where is liberation? When liberation has disappeared, there is nothing at all. Therefore, O fortunate Goddess, the body is to be vigorously preserved.

The venerable Goddess spoke:

17 The true nature of liberation in the body has now been heard by me. If there is compassion for me [in thy heart, then] tell me the step by step means to success in that.

The venerable Bhairava spoke:

18 The preservation of the body is realized through the Work (*karmayoga*). Mercury and breath [control comprise] the Work in two parts.

19 When swooning, mercury, like the breath, carries off disease; when killed, it raises from the dead; when bound, it affords the power of flight.

20 Liberation arises from higher knowledge, [and] higher knowledge from the preservation of the vital breaths. Therefore, when there is stability, mercury is empowered and the body is stabilized.

21 O Goddess! After a short time, the mercurial Work generates immortality in an unageing body, in much the same way that the workings of the mind generate thoughts.

22 Truly, O Goddess, the person who consumes killed mercury obtains the knowledge of higher things together with worldly knowledge, and his mantras become effective.

23 If, however, there is no descent of the Goddess (*śaktipāta*), and if the fetter (*pāśa*) [that binds him] to this world remains uncut, [one must ask,] "How can he have the mental clarity [to proceed] once his mercury has been killed?"

24 For those persons who have completely lost their minds through their chronic addiction to liquor, flesh, vulvas, and penises, higher alchemical knowledge is exceedingly difficult to realize.

25 Furthermore, concerning persons who are devoid of the clan teachings [and] who are without desire for the true doctrine: their mercury is not perfected. They, O Goddess, are [like men] drinking a mirage.

26 He who eats beef and drinks the liquor of immortality I consider to be a member of the clan and a connoisseur of *rasa*. [All] others are inferior.

27 [To those who say] "There is no content [here] in the sense of an established doctrine," it is maintained that *rasa* is that content. It is by means of it that *siddhi* is obtained. Without mercury (*sūtaka*) there can be no *siddhi*.

28 Until such time as a person eats Śiva's seed—that is, his *rasa*, mercury, [one must ask,] "Where is his liberation, where is the preservation of his body?"

29–30 Ignorant persons who are completely besotted with liquor and flesh and deluded by Śiva's illusion will prattle that, "We have been liberated, we have gone to the house of Śiva"—and so those dimwits will not long for the Work that is the preservation of the body. The world, O Mistress of the Gods, is fond of fractured factoids!

31 My darling! One should repeatedly apply one's higher alchemical knowledge in such a way as to produce the power of embodied flight and identity with Śiva.

The venerable Goddess spoke:

32 I wish to hear of the worldly descent and the magnificence of mercury, O Lord of the Gods! Thou canst accurately declare that.

The venerable Bhairava spoke:

33 Excellent, O fortunate Goddess! That which was requested by thee is the most secret of secrets, a gracious consideration for the welfare of the world of men.

34 Thou, O Goddess, art the mother of all beings and I am their eternal father. That which was generated from our great sexual union is [called] *rasa*.

35 O Goddess! Due to its unconstrained capacity, mercury (*pārada*) is praised for its greatness, and that which is called *pārada* by the best of practitioners is [so called because it affords (-*da*)] the highest (*pāra*) advantage.

36 Mercury (*sūta*) is identical to me, o Goddess. It arises from my every limb. It is my bodily fluid (*deharasa*), which is why it is called *rasa*.

37–38 By viewing, touching, eating, indeed by simply recalling, worshiping, and offering mercury, its six-fold reward is made manifest. The merit [gained] from viewing all the liṅgams in the world—Kedārnāth and so on—as well as all others that may exist, is [realized] through the simple viewing of mercury.

39 Mercury that has become "swooned" (*mūrcchita*) when compounded with sandal, camphor, and saffron [is employed] in the worship of Śiva, for the *siddhi* of [eternal] nearness to Śiva.[31]

40 Through its ingestion, O Supreme Mistress, mercury slays the triad of sins. It likewise slays the three afflictions, and [halts the] generation of the [three] morbific entities (*doṣas*).[32]

41–42 The highest station, difficult even for learned brahmins to reach, is attained. By calling to mind the Champion of Rasas (*rasendra*) situated in the center of the lotus in the space within one's heart, O Supreme Mistress, one is immediately freed from the accumulated sins of former lives.

43 The combined rewards [gained] from worshiping 1,000 of Śiva's self-generated (*svayambhū*) liṅgams are increased by a factor of ten million by worshiping a [single] mercurial liṅgam.

44 Now, the "excavational doctrine" is the lowest [alchemical doctrine] and the "portal (or pit) doctrine"[33] is middling, but the "mantra doctrine" is the highest [and] the "mercurial doctrine" is the very highest.

[31] This is the goal of devotional Śaivism. See above, note 1.
[32] These three—bile, phlegm, and wind—are the three humours of Ayurveda, for which humoral imbalances are the prime source of disease.
[33] See below, RA 2.9–11, and above, notes 23, 25.

45 Those who corrupt the alchemical Work on the basis of their expertise in sorcery and spells all go to straight to hell, having destroyed the accumulated good acts [of their former lives].

46 The mercurial science is the supreme science, exceedingly difficult to attain in the three worlds. Because it affords worldly enjoyment and liberation, it should be promulgated by persons eminent in the field.

47–48 Some say "it (i.e., the mercurial science) exists"; some say "it does not exist." *Siddhi* will accrue to the person who believes it exists: he realizes his worldly goals. My darling! [As for] the person who says, by way of indicating his disbelief, that "it does not exist, it does not exist," his *siddhi* will not exist, not even after a billion rebirths.

49–52 The sinner who disparages mercury, considering himself to be liberated through [mere] knowledge of the absolute: he loses my protection for even a billion rebirths. He is born as a dog, o Goddess, for a full thousand births. [As for] the sinner who disparages mercury: due to that [sinful] disposition, he is born as a cat for 300 trillion births, a donkey for 100,000 births, a crow for 100,000 births, a worm for 100,000 births, a junglefowl for 100,000 births, and a vulture for 100,000 births. He who converses or has bodily contact with revilers of mercury is also burdened with sorrow for up to a thousand rebirths.

53 Moreover, on account of its fluidity, potency, and power of digestion,[34] mercury is the supreme ambrosia. By virtue of this, my darling, mercury wards off death and rebirth, disease and injury.

54 To begin, a person should, with purified inner soul, propitiate his guru. Once the guru has been satisfied, the rays of light emanating from his teachings shall impart the established doctrine.

55 The dimwit who undertakes the alchemical Work without serving a guru meets with no success. His Work is as worthless as treasure discovered in dreams.

[34] Here, the medical term *vipāka* ("digestion") refers to mercury's ability to swallow or assimilate other reagents in the series of alchemical processes known as *jāraṇa*, the thirteenth alchemical *saṃskāra*.

56 He who undertakes the valid Work meets with success at every step, and so, my dear, his guru, once satisfied, should cause him to put the mercurial doctrine into practice.

57–58 This teaching that is a path to *siddhi* affords both worldly pleasures and liberation. This great Tantra called the *Ocean of Mercury*, although it be exceedingly rare and esoteric, may be accessed, for the realization of its benefits, through the grace of the guru. In this regard, one should not, having succeeded in the alchemical Work, behave selfishly.

59 Then, having been instructed by guru, and having received the [Goddess's] Command[35] and taken recourse to Bhairavī embodied as the Elephant Goad of Mercury (Rasāṅkuśī), one should undertake the worship of the Mercurial Bhairava.

60 O Mistress of the Gods! The origin and greatness of mercury have thus been declared. Tell me, O Goddess, what else dost thou wish to hear?

Thus [ends] the first chapter, named "The Descent of the Tantra" in the alchemical compendium [titled] the "Flood of Mercury" [as elucidated] in the dialogue between the Daughter of the Mountain (Pārvatī) and the Supreme Master (Parameśvara).

Rasārṇava, Chapter 2

The venerable Goddess spoke:

1 Please tell me [this], O Lord! What sort [of person] should the alchemical preceptor be? And, O God, what so with the apprentice who makes the practice of alchemy his supreme goal?

The venerable Bhairava spoke:

2–6 [Such a person] is considered [to be a] guru: [he is] free from desire, egoless, devoid of avarice or duplicity, devoted to the path of the Clan (*kula-marga*), ever devoted to the veneration of his [own]

[35] On the technical sense of this term in the Tantras of the Western Kaula Tradition (*paścimāmnāya*), which identify *ājñā* as both fluid medium (in the form of the goddess Kubjikā's sexual emission) and esoteric message, see Dyczkowski 2004: 177–179.

guru, self-restrained, pedagogically savvy, energetic, unselfish, upright, truthful, dexterous, good-natured, possessed of every good quality, pure, knowledgeable in multiple alchemical teachings, possessed of laboratory experience, [and] well-versed in the precepts of alchemical initiation. He is [also] someone who has full knowledge of apparatus, botanicals, the primary reagents, the nuances of color, the components of the mineral seeds (*bījas*), the conjoining of opposites, calcinating salt, coloring, flowing, oil, metallic "leaf" preparations, progression, the extraction of color and pliability, [and] the "youthful" and "aged" as well as the "sky-going" and "earth-going" [types of] calcination.[36]

7–8 The [prospective] apprentice is knowledgeable about what is to be done according to place and time. [He is] compassionate and dexterous, entirely free of avarice and duplicity, having mantra practice as his highest goal, endowed with [auspicious] bodily markings and attributes[37] [and] depth [of character], beloved by his teacher, ever devoted to the veneration of the gods, to fire, to the circle of Yoginīs, and to the Clan, a cultivated connoisseur of Tantra, a speaker of truth, and firm in his vows.

9 Men who are equipped with[38] hands that bear the marks[39] of water-pots, spades, banners, conches, and so forth are, O Goddess, to be employed in "treasure" (*nidhi-*) practice.

10 [Men who are] ever ferocious, powerful, [and] massive brutes with dark bloodshot eyes and bent noses are recommended for "pit" (*vila-*) practice.

11 In smelting (*dhātuvāda*), one should employ vigorous red-headed [men] with emaciated upper bodies,[40] and hot and hardened feet.

[36] I have emended *jaranāṃ balabaddhayoḥ* to read *jāraṇā-balavṛddhaye*, a neuter accusative dual signifying "youthful and aged calcination," that is, the stadial consumption of ever-increasing quantities of mica or sulfur by mercury, until said mercury becomes *baddha*, fixed. "Coloring," "flowing," "progression," and "calcination" comprise the fourteenth, fifteenth, sixteenth, and thirteenth of the standard set of eighteen alchemical processes (*saṃskāras*), respectively. These immediately precede the final two processes—or results—of the prior *saṃskāras*: transmutation and bodily transformation. The fields and procedures of which the guru has "full knowledge" conform closely to the themes of the *Rasārṇava*'s chapters 3–16.

[37] Here, the compound *sāmudralakṣaṇopeta* is to be understood in the context of *sāmudrikaśāstra*, the "science of physiognomy": Guenzi 2021: 127.

[38] Troubling here is the verbal form *adhi-ṣṭhita*, which would normally mean "settled" or "inhabited."

[39] Here I have adopted the Kashmir manuscript's reading of *lakṣaṇaiḥ* in place of *lāṅchitaiḥ* as given in the critical readings of the two editions used here.

[40] My reading for the problematic *ūrddh[v]apiṇḍikān*.

12 At the outset, O Goddess, [the guru] should put to the test well-composed brahmin, kṣatriya, vaiśya, and śūdra initiands, in order [of caste].

13 [Persons who are] in control of their senses, inured to pain, persevering, valiant, and of consummate wisdom are considered to be good initiands.

14 [However, the guru] should vigorously reject persons without serious preparation in matters alchemical, who are lacking in the aforementioned qualities, [and] who are dissolute vagabonds.

15 One should completely shun women who are evildoers, ill-behaved, cruel, quarrelsome, and gluttonous, and [women who are] melancholic, with hollowed eyes, as well as [women who are] without mercy and who speak ill of mercury.

16 The earth's disk is filled with such people, O fair-hipped one! A woman of this type is not [to be employed as] a "fixa-trix of mercury" (*rasabandhakī*).

17 [There is] the "cowrie-shell" (*kākiṇī*) [and] the *kīkaṇī* woman, as well as the "glass-bead girdled" (*kāñcikācinī*) [female type]. The woman who comes into season in the moon's dark fort-night is styled a "cowrie-shell" woman.

18 She who comes into season on both fortnights is styled a *kīkaṇī* woman.

19 She who comes into season on the bright fortnight is considered [to be] a "glass-bead girdled" woman.

20–25a Now, a woman who is in the full bloom of youth, fair of form, of pleasant laughter [and] lovely hair, a perennial lover of dairy products, agreeable in her conversation, [and] ever fond of the tales and teachings of Śiva; whose mouth is lotus-shaped [and each of] whose eye[s] has the appearance of a blue lotus flower, whose teeth are like diamonds, whose lower lip resembles a tender shoot, whose breasts are even, prominent and full; and whose vulva—the two parts[41] of which are fleshy, smooth, [and] rounded, with inward curling black pubic hair—is symmet-rical and resembling a fig leaf; and who is slim at the waist and

[41] Presumably, the *labia majora* and *labia minora*.

supreme in [her] devotion to Śiva: [such a] slender, gracious, and doe-eyed woman who menstruates on the full moon or the new moon, from fortnight to fortnight, is to be known, assuredly, as a "lotus woman."

25b–26b[42] The aforementioned [woman] called "cowrie shell" is of six types. She is a provider of the essential fluid. O Mistress of the God of Gods, [it is] with her, precisely, that one should perform the alchemical procedures.

27 The supernatural powers [obtained] on the alchemical path belong to him in whom the Great God (Mahādeva) is well pleased.[43] Surely, that person is of clear purpose, and resolute in his alchemical practice,[44] and for that reason the guru and the aforementioned initiand [each] becomes a co-sharer in [those] supernatural powers.

The venerable Goddess spoke:

28–29 Where does a practitioner find a woman with such distinctive features? Here [in this world], where [there is] beauty, O Great God, the distinctive feature is not found.[45] [And, in cases] where the distinctive feature is present, there's no enthusiasm. Put another way, how does beauty manifest in an unattractive woman?

30 Where has the [optimal] triad of distinctive features been seen by thee, O Great Master? And on that subject, tell me, O Lord, the means by which that [triad] is generated [in a woman].

The venerable Bhairava spoke:

31 Listen, O Goddess to something that is a supreme mystery, difficult to fathom even by the gods, [i.e.,] a life engendered by Sadāśiva in the belly of Caṇḍikā.

[42] Lines 26a and 26b are reversed in the Ray and Kaviratna edition. They are in the correct order in Tripathi's edition.

[43] This hemistich is absent from the Ray and Kaviratna edition.

[44] *Tasya hi nirmalā buddhirniścintā rasasādhane*: nearly identical language is used below, in 2.32a, in the description of the alchemist's female assistants (*tāsāṃ buddhirbhaveddevi nirmalā rasasādhane*)

[45] Presumably, the ability to menstruate in accordance with the lunar cycles, and, in the case of the *kīkaṇī*, menstruating twice a month. This, together with beauty and enthusiasm, appear to constitute the *Rasārṇava*'s triad of distinctive features (*lakṣaṇa-tritayam*), rather than the better-known Buddhist *trilakṣaṇa* (impermanence, unsatisfactoriness, suffering) or something more Socratic (goodness, truth, beauty).

32 [In the case] of those [women who are to be employed] in the alchemical practice, virtuous intent arises [in the following way]: One shall cause [the] Expeditious[46] [spell] to be given, [and] a woman who is without arrogance shall incant that mantra.

33 She who would incant the mantra 100,000 [times] shall be (re) born as a woman possessed of the optimal distinctive feature.[47] Next, making the Aviator's seal (*khecarī-mudrā*),[48] she should incant the Howling Bhairavī [mantra].

34 Next, she shall recall [the following]: "This, [the goddess] Caṇḍikā is my mother, and [my] birth is from the Terminator of the Triple City [i.e., Śiva]. I am the arisen Kālikā. Such [a person] am I."

35 Next, she shall declare past, and future in the triplicate cosmos. [In the end,] the quality of one's female assistant [will accord with] the quality of one's (i.e., the alchemist's) destiny.

The venerable Goddess spoke:

36 O god of gods! O Great God! O embodiment of all wisdom! For the welfare of practitioners, I now inquire [about] alchemical initiation.

The venerable Bhairava spoke:

37 Listen, O Bhairavī [as] I describe to thee something entirely new: the performance of alchemical initiation. Therefore, listen to what is proclaimed.

38–39 [In a country] where the king is powerful, entirely devoted to religion, and having good governance as his highest aim, and [where the royal] minister is a connoisseur of the true nature of religion [and] entirely possessed of devotion and faith: it is there that one should perform the Work, [i.e.,] the alchemical path [that consists] of the fixation of mercury. One should practice in a country where the people are sectarian devotees of the Great Lord (i.e., Śiva).

46 The *tvaritā* mantra, identified with the great goddess Durgā.

47 As an adjective, *sulakṣaṇā* can denote either the single "feature" of menstruating according to lunar cycles, or the triad alluded to in note 46.

48 The hathayogic *khecarī mudrā* involves concentrating the gaze on a point between the eyebrows while pushing the tongue upward into an opening at the back of the soft palate (White 1996: 253–254). With the tongue in that position, it would be difficult to actually incant a mantra, so perhaps the verb (*japet*) should be understood as denoting its silent pronunciation.

40 One should practice in an uninhabited place where no domestic animals are found, on a well-secluded, level site free of any sort of encumbrance.

41–43 O Supreme Mistress! One should practice [alchemy] on a pleasant, unsullied, and isolated site, in a decorated house whose well-watered garden is equipped with a moat and a surrounding wall secured by a bolted door. Abounding in various flowers and trees [and] dotted with an assortment of geese and ducks, [and further] beautified by [the presence of] *brahmany* ducks, [the garden] is graced with a profusion of celestial herbs, sweet-smelling green grasses, red lotuses, blue lotus blossoms, and white esculent water lilies, and adorned with groves of plantain trees.

44–47a In the center of that [garden], O Mistress of the God of Gods, one should see to the construction of a one- or two-storied hall fitted out with windows [and] a raised exhaust vent for smoke. [It should be] hung with red banners and equipped with [alchemical] apparatus, [as well as] a duly independent store-house for assorted varieties of herbs, grain chaff, wood, and cow dung.[49] [It should have] the god of the place[50] painted [on one of its walls and] a Remover [of Obstacles[51]] venerated [at its entrance]. [The hall] is sanctified by [images of] Umā and Īśa, [and] guarded by the [eight] World Protectors.

47b–48b Now, O Goddess, on a fortnight when the moon and stars are brightly shining, on an auspicious lunar date, and at the precise moment when an auspicious planet is beheld on the horizon, the freshly bathed [initiand] is clothed in white

[49] The text gives *āraṇya-utpala* here, which can signify either "wild crepe ginger" or the "wild blue lotus blossom," neither of which is a satisfactory reading, coming at the end of what is most likely a list of combustibles used for the controlled heating of various alchemical preparations. I therefore suggest the reading *āraṇya-upala*—"cow dung" as an alternative to the more typical term *vanopala*, which denotes the dung of cows found in the woods. See Hellwig 2009: 218 on *vanopala*.

[50] It is unusual to find the term *digdeva* in the singular: the directional guardians are nearly always referred to in the plural, because they guard the eight cardinal and secondary directions. As such the plural term is synonymous with *lokapālāḥ* ("World Guardians"), which appears in 2.46a. The names of these regents of the eight directions are given in full in 2.107–108.

[51] Given here in the singular, *vināyaka* is probably synonymous with Gaṇeśa, the elephant-headed god whose image is frequently found guarding the entrance to a temple. For the history of the multiple *vinayakas* and their historical relationship to the Vighnas, Gaṇapatis, and Gaṇeśa, see Thapan 1997: 91–97.

garments and anointed with unguents derived from white flowers.

49a–52a After he has worshiped maidens with honey, clarified butter, yoghurt, milk, and sesame seeds, O Mistress of the Gods, let him construct a brick altar measuring ten hand-breadths on each side [and] smeared all over with the dung of a brown cow and surrounded by golden pots strewn with barley and white mustard seeds and made beautiful with fragrant garlands. Then, after he has laid a stone slab over it and deposited the slab's [magic] bolt[52] onto his fingers, heart, and so forth with the "Elephant Goad of Mercury" [mantra], he should, my darling, install a mercurial *liṅgam* there [atop the slab], together with gold.[53]

52b–53a "*Oṃ hrīṃ śrīṃ*" is declared [to be] the root mantra of King Mercury (*sūtarāja*). Using the root mantra, one should invite the goddess [named] "She Who is the Elephant Goad of Mercury" [to descend] onto that [*liṅgam*].

53b–54b And worshiping the directional guardians on the four edges [of an alchemical maṇḍala], he should install in order, on the outside, [the following]: Nandin, Mahākāla, Bhṛṅgirīṭa, Mahābala, and Kumbhakarṇa, Sugrīva, Bhṛṅgīka, and Dṛḍhāyudha.

55a Having deposited [them] fourfold—[that is,] in groups of four at the gateway[s of the *maṇḍala*]—he should worship [them] on the right and left [doorposts of each gateway].

55b–56b [Now moving inside the square borders of the *maṇḍala*,] having worshiped Śakra on the eastern [quadrant],[54] he should deposit Skanda in the northeast, and thereafter Rudra in the south and likewise Vāyu in the southwest. Now, [he

[52] My reading of the problematic *śilāpaṭṭam* is tentative here. It is difficult to imagine a bolt made of stone securing a stone slab to a brick altar. Since this is the first of many verses that bring mantras into play, I have chosen to interpret this as a bolt-mantra, a magical seal incorporated into the alchemist's body to secure the altar that will serve as the support for the elaborate *maṇḍala* described in the verses that follow.

[53] The use and placement of the gold referred to in this passage are unclear. Like all tantric *liṅgams*, the mercurial *liṅgam* evoked here is set into a chasing, called a *yoni* ("vulva"); a portion of a silver ring, which likely served this purpose, was excavated together with a mercurial *liṅgam* at the Chandi Bukit Batu Pahat temple site in Malaysia. On this, see Treloar 1972: 237–238.

[54] Here, the description returns to the placement of the directional guardians evoked in 2.53b, who are situated on the edges, as opposed to the outer gates, of the *maṇḍala*'s square *bhūpura* enclosure.

should deposit] Śiva in the western area, Agni in the northwest, and Umā in the northern area. [One should deposit] Vyāpaka (the "Emanator") in the northeast.[55]

57–58 The Mistress of Smearing, Mistress of Projecting, Mistress of Corrosion, and likewise the Mistress of Coloration, Metallica, and Fixatrix, She Who Moves on Land, She Who is the Destroyer of Death, the Mistress of Superhuman Powers as well as She Who Flies through the Air[56]—and, in sequence, the ten Female Messengers (*dūtīs*): these [two sets of ten superhuman female beings] are to be worshiped on an eight-petalled lotus, both above and below [?the lotus's calyx] as well as on its petals.

59 Copper pyrite, iron pyrite, bitumen, selenium, calamine, and likewise copper sulfate, sulfur and orpiment [are also to be worshiped] sequentially [on another set of eight inner petals] starting from the east.

60–61 Likewise, all of the eight Masters of Wisdom (*vidyeśvaras*)—eighteen-armed, dreadful, five faced, three-eyed, wearing diadems adorned with the half moon, blue throated, bull-bannered, each bearing his own distinctive color—are to be worshiped, O Great Mistress, on [the petals of] another [inner] eight-petaled lotus.

62 On that [innermost] lotus's pericarp, one should first worship the group of four Energies (*śakti-catuṣṭayam*), starting in the east. Now, Mālinī is the energy of gold, Parā is termed the energy of mica,[57] Aparā is [the energy of] diamond [while] Parāparā is to be known as [the energy of] magnetite.

63–65 Now, [seated] in the midst of those [four] Energies is the omniscient Rasabhairava who has the appearance of pure crystal, [and] five faces, [each with] three eyes. [His] terrible eyes are flaming yellow, and likewise his mouth and tongue

[55] These names differ significantly from standard nomenclature. Moving clockwise from the east, these are usually named Indra (= Śakra), Agni, Yama, Nirṛti, Varuṇa, Vāyu, Kubera, and Iśāna.

[56] Lepikā, Kṣepikā, Kṣārikā, Rañjikā, Lohaṭī, Bandhakārī, Bhūcarī, Mṛtyunāśinī, Vibhūti, and Khecarī.

[57] Here, I have emended *balā* (Sida cordifolia), clearly a scribal error, with the term that the *Rasendracūḍāmṇi* (2.24) uses in its place: *abhra*, "mica." As a mineral form of the Goddess's sexual emission, this is likely the correct reading. The *Rasendracūḍāmaṇi* follows the *Rasārṇava*'s lead for the names of the other three mineral energies.

are aflame, his brows are aflame, [and his] hair is aflame. He is difficult to behold. His throne is a ghost. He sports a crown of matted hair and has the half-moon for his diadem. He wears the skin of a tiger, has a serpent for his sacred thread, and a bull on his banner.

66–67 One should worship this eighteen-armed god whose mouth is faintly smiling[58] [and] who is like Śiva in all his parts, with the thirty-two-syllabled mantra. So listen to me as I enunciate the mantra of the Lord of Mercury (Raseśvara):

68 Oṃ hrīṃ krīṃ:[59] O Lord of Mercury! O Great Black One! O Very Powerful One! O Aghora ("Unterrible") Bhairava! O Adamantine Virile Hero! O Skeleton of Fury! Kṣlauḥ! Kṣlaḥ!

69–71 In his [Rasabhairava's] sexual embrace is the great Goddess, adorned with jewels and finery, resplendent as molten gold, [and wearing] yellow garments. [She is] three-eyed, framed by a pair of white yak-tails, illuminated by her opened parasol, holding an elephant goad and rosary in her two right hands and a noose and [displaying] the "fear not" [*mudrā*] in her two left [hands]. Wearing the half-moon for her diadem, the blue-throated great goddess Rasāṅkuśī is the embodiment of compassion.

72a–73a [The alchemist] should, for the sake of the supernatural powers accruing to mercury, worship [her] with the spell (*vidyā*) comprised of five seed-syllables. O Mistress of the Gods! The guru should dictate to the initiand the *vidyā* [given here in code as follows:] the Origin of Speech plus the Mistress of the World plus the Seed-[mantra] of Sovereignty, the Maddening One and the Seed-[mantra] of Energy.[60] This *vidyā* is extremely rare.

73b–76b Worshiping according to precept with incense, oil lamps, edible food offerings, flowers, betel leaf, sandalwood, together with the intoning of the "Recital for Pacification" (*śāntipāṭha*)

[58] One of Bhairava's standard attributes, according to manuals of iconography and as seen on his imagery throughout Nepal, is his partly opened mouth, which reveals—but only slightly—his set of fanged teeth.

[59] The correct mantra of Svacchandabhairava contains *kroṃ* in the place of *krīṃ*, which is the reading given here: personal communication from Alexis Sanderson, July 6, 1992.

[60] A *vidyā* is a female spell-deity, the counterpart to a male *mantra*. Decoded, the five seed syllables are: *aiṃ*, *hrīṃ*, *śrīṃ*, *klīṃ*, and *hrī*.

[and] the sounds of hymns of praise and benediction accompanied by clanging bells, he should energetically offer tribute (*bali*) for the pacification of all of the Obstructers—[that is,] for the Trespassed, the Yakṣas and Guardians, and the Flesh-Eaters—with the "Unterrible" [mantra], which begins with "Oṃ" and ends with "obeisance" (*namaḥ*).[61] It is the mother lode of every alchemical Work, [and,] O Goddess, the destruction of [all] calamities and Obstructers.

77a–79b Now, declaiming the "Elephant Goad of Mercury" *vidyā* with all his might, [and], O Mistress of the Gods, preparing the fire pit [comprising] a rounded hole with girdled lines drawn around its opening, [and] making the fire offering (*homa*) in proper sequence with clarified butter and sesame seeds,[62] the [guru who is a] connoisseur of mantras should, O goddess, have the [consecration] pot filled with fluids, topped off with fruits, wrapped in bolts of cloth [and] fitted out with the five jewels.[63] Then [the pot] is consecrated with the eighteen *vidyās*, [incanted] one by one.

80a–81a [Next], venerating the vessel containing the *arghya* water[64] (a vessel filled with fragrant flowers and other essences and adorned with sprays of flower blossoms), he should sprinkle the initiand 108 times with water from the *arghya* vessel, water that has been incanted with the "Granter of Increase" [*vidyā*].[65]

81b–82b And now, having properly consecrated [him] with the water from the pot, he should, O Goddess, dictate the *vidyā* [to him and] have the alchemical initiand recite [it] to virgins, *yoginīs*, yogis, and practitioners, in the usual way.

[61] *Oṃ huṃ aghorahṛdayāya namaḥ.*

[62] Ritual fire pits and fie-pots are surrounded by cords, or have lines drawn around them, called "girdles" (*mekhalas*). This fire offering (*homa*) is a dramatic one, with clarified butter creating a sheet of flame and sesame seeds a cacophony of small explosions.

[63] Gold, diamond, sapphire, ruby, pearl.

[64] Prior to the advent of Tantra, *arghya* denoted the water ritually offered to an honored human guest: in the *Mahābhārata*, the offering of the *arghya* to Kṛṣṇa at Yudhiṣṭhira's royal consecration triggered the schism within the House of Kuru that led to the great war. In tantric parlance, however, this specially prepared water is offered in homage to Śiva or some other divinity: Brunner, Oberhammer, and Padoux 2000: 140, s.v. "arghya."

[65] In his Hindi-language translation of this verse, Tripathi reads "Granter of Increase" (*vardhanyā*) as a hand gesture (*mudrā*) rather than as a *vidyā*.

82b–83a And he should have [the initiand] satisfy [them] with food and drink, and he should have [him] observe a night vigil there.

83b–85b O heroine of the gods! Thus the practitioner whose initiation has been completed, should, having bound the "Elephant Goad of Mercury," seal (*rasāṅkuśī mudrā*) onto a thread whose essence is breath control, recite 100,000 times the five-seed [-syllabled] *vidyā*,[66] which, conjoined with the set of five of sounds, is comprised of the five elements, my darling! Unimpeded [identity with] Rasabhairava is realized by him, O Mistress of the Gods!

86–88 The supreme utterance that begins with Oṃ and ends with "obeisance" (i.e., the "Unterrible" mantra) [is to be recited] at the conclusion of the water offering. O Mistress of the Gods! Gathering together the sandalwood of non-injury, the flower of truthfulness, the incense of not stealing, the sexual moderation that is the great lamp, the non-acceptance of gifts that is the little bell, and the dairy-based food and drink whose essence is compassion toward all beings, he should propitiate the Goddess. He [thereby] partakes of the supernatural powers. He should recite [this mantra] to her either 1,000 or 100 [times] at dawn, noon, and dusk.

89 O Thou who art praised by the gods! By the grace of her Command, sky-faring mercury arises,[67] and divine botanicals fall to his share in the same manner.

90 Just like a person who, mounting a mad elephant in the forest, [finds himself] without an elephant goad, so [it is for] the wily one unversed in the occult science of the "Elephant Goad of Mercury": gaining hold of mercury, he perishes.

91 [Equipped] with the *vidyā*, the person who fully desires the supernatural powers that arise from the use of mercury should set his mind on a guru belonging to an authentic lineage.

92 Hereupon, in order to bring about the descent of [the goddess] Praśnā, [the newly initiated alchemist] should, with mind

[66] See above, note 60.

[67] Mercury that has reached an advanced state of perfection becomes *khecara*, possessed of the power of flight. The alchemist who inserts a pill (*guṭikā*) of said mercury into his soft palate is then himself empowered to fly through the air and sport with airborne demigods and goddesses: *Rasārṇava* 12.336–337.

composed, meditate on the aforementioned Rasabhairava. He should then proceed in such a way as to dwell within [that god].

93 Fixing the mind on the unstruck [sound] in the fontanel, he should cause it to cease.[68] Having first effected the depositing of [the mantra deities on] the hand [and], without pause, the depositing of [the mantra deities on] the limbs of the body, he should, in his meditative state, declaim the mantra of the Champion of Rasas with all his might.

94 Making a square once more, he should, using spotless grains of rice, trace out a hexagonal *maṇḍala*, either one or two hands in size, inside [that square].

95 On that spot besmeared with sweet-smelling substances, he should worship [Rasendra] with [offerings of] sandalwood [paste] and other [fragrant essences].

96 On the calyx [of the lotus at the center of the maṇḍala], O Goddess, he should deposit the aforementioned Rasabhairava. On the six corners [of the hexagonal *maṇḍala*] he should deposit the hexad of deities, beginning with Mahākāla.

97 He should deposit Mahākāla, Mahābala, Aghora, Vajravīraka, Krodha, and Kaṅkāla, [and] honor [them] with the garland mantras.

"OṂ hrīṃ kṣlaiṃ kṣlaṃ hriṃ hrīṃ hrīṃ hrūḥ huṃ phaṭ! [Obeisance to] Mahākāla Bhairava, the Lord of Mercury! [Obeisance to] Him of fierce appearance, whose eyes are bile-injected and black. Descend! descend! cause to descend! cause to descend! Speak, speak! cause to speak! cause to speak! Declare the pure and impure! declare the pure and impure! Cause to declare! cause to declare![69] Provide optimal protection! provide optimal protection! Bestow the supernatural power of mercury! bestow the supernatural power of mercury!" Such are the garland mantras.

[68] In hathayogic practice, the unstruck (*anāhata*) sound begins to reverberate in the heart, and gradually fades away to silence as the practitioner raises his breath, energy, and mental apparatus into the cranial vault.

[69] The pairing of imperatives with causative imperatives here indicate that it is necessary that the alchemical Bhairava first descend into the maṇḍala in order to trigger the subsequent descent of Praśnā (*avatara* and *avatarāya*), and that the god's intervention is needed to trigger the divinatory speech that will emerge from the mouths of children in the rite (*kathaya* and *kathāpaya*).

"Oṃ, Mahākālabhairava for the heart, *namaḥ*! Oṃ Mahābalabhairava on the head, *svāhā*! Oṃ Aghorabhairava for the topknot, *vauṣaṭ*! Oṃ Vajravīrabhairava for the amulet, *huṃ*! Oṃ Krodhabhairava, for the two eyes, *vauṣaṭ*! Oṃ Kaṅkālabhairava for the missile, *phaṭ*!" In every case, *svāhā* [comes] at the conclusion [of the utterance]. [It is] in this way [that] acts of depositing on the body [are to be done]. In this way, one should first perform the depositing [of the garland mantras] on the fingers. When worshiping these six limbs, thc root mantras [are employed].

98 The depositing is [performed] thusly with the clustered syllables [of the mantras]. He should worship [these mantra deities] with fragrant flowers. He should place a [water-]filled pot over the center,[70] together with a platter [whose surface is scattered] with grains of rice.

99 On top of that [he should place] an oil lamp [and] a wick, and he should incant [that lamp] with mantras.

100 Having brought together a girl and a handsome boy, he should assemble one, two, three, four, or five [male or female children]—according to however many are available—and he should venerate [them] with offerings of fragrant flowers, incense, oil lamps, and edibles.

101 He should deposit the mantra deities, beginning with [the deity of] the heart, on each of those girls.

102 Then, having gazed into that lamp, all of the girls [who are present] there [will] perceive deities in its flame, and the boys [will see] both good and evil. The girls [will] declare either success or failure [in the alchemical Work].

103 Understanding this to be the descent of Praśnā, [the newly initiated alchemist may now] undertake the alchemical Work.

104 O Lord of the God of Gods! A person whose soul who was formerly (i.e., in past lives) focused on Rasendra—[that is, a person] who has had seven rebirths, [i.e.,] who was born and died [six previous times]—should not abandon his enthusiasm for mercury [in this life].

[70] Presumably, the center of the lotus calyx mentioned in 2.96.

105–106 Thus knowing the evil and the good, a person who has obtained the local deity's favor should—after worshiping Rasāṅkuśī in the laboratory as before, and after greatly honoring the preceptor with incense, garlands, sandalwood [paste], and so forth, and offering tribute with the "Unterrible" [mantra]—now begin the Work.

107–108 In the laboratory, botanicals [are to be installed] in the east, mercury condensing [apparatus] in Agni's range (southeast), metal-killing [reagents] in the south, and grinding [apparatus] in the Nairṛti quadrant (southwest). Now, liquefaction processes [are to be carried out] in Varuṇa's quarter (west). The bellows [are to be installed] in the direction of Vāyu (northwest), my dear, and the extraction of color [is to be undertaken] in the direction of Kubera (north), [and] the transmutation of mercury in the direction of Īśāna (northeast).

109 After installing the guru's throne in the center [of the laboratory], O heroine of the gods, the Work that begins with subduing and ends with coursing,[71] O fair-faced one, is to be carried out by the [newly initiated] practitioner with the "Elephant Goad of Mercury" mantra.

110–111 [Concerning] Rasabhairava, who was previously described, O Mistress of the God of Gods, his knee is Sadyojāta, and Vāmadeva his genitalia. Aghora is his heart, and his mouth is declared [to be] Tatpuruṣa. Extending up to the middle of his eyebrows is Īśāna, [who wears] the half-moon [on his] brow.

112 The "drop" is above that; placed above the drop is the "vibration." [The area] between the forehead and the top of the head is the very place where Energy resides.

113a–114a She Who Pervades is situated on the fontanel, and above her She Who is Transmental may be found. [The] Transmental is the state of being beyond mind. The realm of the Transmental is imperishable. Above that is the Satya [realm]. It is situated in the ether, higher than the highest.[72]

[71] These are *niy[ā]mana*, the seventh alchemical *samskāra*, and *krāmaṇa*, the sixteenth *samskāra*, respectively.

[72] This appears to be an imaginative fusion of Puranic cosmology—in which the realm of Satya ("Truth") is the highest of the seven worlds inside the cosmic egg—and Samkhyan metaphysics,

114b–116b He who is empty, the void, the highest void, and the entire triple void, as well as the sky, firmament, space, vacuity, and the ether; He who is absolute, non-particulate, immaculate, eternal, motionless, infallible, immaterial, unsupported, [and] the attributeless abode of the attributes: one should always meditate on Rasabhairava in this form within one's own body.

117–118 Now, when one meditates unwaveringly [on Rasabhairava] and when one's mind is steady, then mercury, which is likewise steady, is fixed in the midst of fire. But when one's thought wanders, mercury does not remain stable in fire.

119 By night, [the newly initiated alchemist] should worship the Dryads, Gandharvas,[73] Flesh-Eaters, and likewise the Guardians on the outside of the maṇḍala with alcohol, fish, flesh, and so forth.

120 [Other] Obstructors to the doing of what is to be done and Obstructors to the fruits [of those acts] likewise exist in the tens of millions; and for the pacification of every one of them, he should, after first worshiping the Master Who Is a Boy (Baṭukeśvara) and the hexad [of gods[74]] whose worship brings favor, accomplish [the Work].

121 Also, at the conclusion of the Work, after again greatly honoring the boy [i.e., Baṭukeśvara] whose grace takes the form of sixty-four [Yoginīs],[75] he may then employ the perfected elixir that affords the supernatural powers.

122 But once [his mercury has been] perfected, he should stay tight-lipped and leave his native land behind.[76] For so long

which identifies ether (*vyoman*) as the highest and most subtle of the five elements. At the same time, two other registers are operative: the macrocosmic body of Rasabhairava (? identified with the guru) and the microcosm of the alchemist's body.

[73] An untranslatable term, referring to heavenly demigods associated with beauty, music, and childbirth. However, in the Vedic period, they are more problematic, portrayed as erstwhile enemies of the gods, dangerous seducers and seductresses, and spirits of the dice, alcohol, and other vices.

[74] Baṭukeśvara is the name found in such coeval tantric and alchemical works as the *Kaulajñānanirṇaya* and the *Kākacaṇḍeśvarīmata*. Baṭuka Bhairava, whose name appears in later tantric and devotional works, may be a later Kaula form of this god. The hexad of gods evoked here is the same as that named above in 2.97.

[75] According to several Kaula traditions, Bhairava in his adult form is the lord of the sixty-four Yoginīs, Yoginīs whose grace (*anugraha*) is necessary for initiation into the Kaula.

[76] Success in alchemy breeds envy and controversy, the greed for gold and immortality being strong incentives to violence against the alchemist. A modern-day case in point concerns Krishna Pal Shastri, who, according to reports, created alchemical gold in 1940 in a laboratory on the campus of Benares Hindu University. According to an eyewitness to that experiment, Shastri was poisoned to death by persons attempting to extract his secrets from him (White 1996: 336–337).

as his great chemical reagent (i.e., mercury) remains fully perfected, he should wander about, [engaging in] bee-like begging.[77]

123 When, due to carelessness, some impediment has arisen during the fixation of mercury, then the pacification [ritual] for mercury is to be performed. One should not perturb mercury.

124 O Goddess! One should not make a show of mercury. One should not let [it] fall into a woman's hand. [This] secret [teaching] is not to be communicated to [any] woman. One should indeed hide the elixir from them.

125 It (i.e., the Work) is not to be carried out with another person's elixirs. One should not touch another person's wives. One should also not exploit others, nor should one confide in them.

126 It (i.e., the elixir) is not to be given to evil-minded persons. One should not bring it to completion outside of one's circle.[78]

127 One should, according to one's means, hold a celebration for mercury on the half moon, full moon, and new moon nights, [as well as] at the beginning of a five-year cycle, on the solstice and the equinox, and especially on the fourteenth day [of each lunar fortnight].

128 O Goddess! Should he devoutly honor mercury in this way, [then his] various and sundry sins—brahmanicide and so forth—will be destroyed.

129 Neither disease-related distress nor the [six] calamities arise in places where practitioners greatly honor [mercury] in this way.[79]

130 After greatly honoring mercury as is appropriate in this way, [the alchemist] who is attentive to his own well-being should then undertake the Work.

[77] The metaphor likens the successful alchemist to the bee that quickly flies from flower to flower to gather tiny amounts of precious pollen. The alchemist must dissemble, and engage in begging to hide the fact that he now possesses a source of inexhaustible wealth.

[78] The "critical" reading of this *pada* is untranslatable. My unsatisfactory translation here is based on the Kashmiri manuscript witness's opaque *goṣṭhībhūy[o] na pācayet*.

[79] The six "canonical" calamities (*ītayaḥ*) are: drought, excessive rain, rat infestation, plague of locusts, parakeet infestation, and foreign invasion.

131 [As for] the person who does otherwise, he does not meet with success. Misfortune and anxiety, even insanity may occur.

132 Contempt, a multitude of afflictions and doubt [will dog him] at every step. On account of [his] sloth and covetousness with regard to his teacher, and as a result of speaking [ill] of another person, O darling, hordes [of malign beings will] steal away his hard-won expertise.

133 He who has been initiated [and] who is conversant in the depositing of mantras should proceed with the Work. Tell me this, O Mistress of the Gods: what other [teachings] dost thou wish to hear?

Thus [ends] the second chapter, named the "Rule for Initiation," in the alchemical compendium [titled] the "Flood of Mercury" [as elucidated] in the dialogue between the Daughter of the Mountain (Pārvatī) and the Supreme Master (Parameśvara).

Bibliography

Ānandakanda of Mahābhairava. See Radhakrishna Shastri 1952.

Balbir, Nalini. 1993. "À la recherche des trésors souterrains." *Journal of the European Āyurvedic Society* 3: 15–55.

Böhtlingk, Otto, ed. and trans. 1865. *Indische Sprüche. Sanskrit und Deutsch*. St. Petersburg: Eggers et Comp.

Brunner, Hélène. 1975. "Le *sādhaka*, personnage oublié du Śivaïsme du sud." *Journal Asiatique* 263 (3–4): 411–443.

Brunner, Hélène. 1986. "Les membres de Śiva." *Asiatische Studien/Études Asiatiques* 40 (2): 89–132.

Brunner, Hélène, Gerard Oberhammer, and André Padoux, eds. 2000. *Tāntrikābhidānakośa I*. Vienna: Österreichischen Akademie de Wissenschaften.

Cakrasaṃvara Tantra. See Gray 2007.

Davis, Richard H. 1991. *Ritual in an Oscillating Universe. Worshiping Śiva in Medieval India*. Princeton, NJ: Princeton University Press.

Dyczkowski, Mark S. J. 2004. *A Journey in the World of the Tantras*. Varanasi: Indica Books.

Gray, David B. 2005. "Eating the Heart of the Brahmin: Representations of Alterity and the Formation of Identity in Tantric Buddhist Discourse. *History of Religions* 45 (1): 45–69.

Gray, David B. 2007. *The Cakrasamvara Tantra (The Discourse of Śrī Heruka) Śrīherukābhidhāna. A Study and Annotated Translation*. New York: American Institute of Buddhist Studies.

Guenzi, Caterina. 2021. *Words of Destiny: Practicing Astrology in North India*. Albany: State University of New York Press.

Hellwig, Oliver. 2009. *Wörterbuch der mittelalterlichen indischen Alchemie*. Groningen: Barkhuis.

Haṭhayogapradīpikā of Svātmarāman. See Iyengar 1972.

Iyengar, Srinivasa, ed. and trans. 1972. *Haṭhayogapradīpikā* of Svātmarāman. Madras: Adyar.

Kathāsaritsāgara of Somadeva. See Shastri 1970.

Misra, Siddhinandan. 1984. *Rasendracūḍāmaṇi by Ācārya Somdeva*. Varanasi: Chowkhambha Orientalia.

Netra Tantra. See Shastri 1926b.

Padoux, André. 1989. *Vāc: The Concept of the Word in Selected Hindu Tantras*. Albany: State University of New York Press.

Radhakrishna Shastri. 1952. *Ānandakanda*, volume 1: *Srirangam: Sri Vilasam Press*. gretil.sub.uni-goettingen.de. Input by Oliver Hellwig.

Rasaratnākara (Ṛddhikhaṇḍa) of Nityanātha. See Shastri 1940.

Rasaratnasamuccaya of Vāgbhaṭṭa II. See Sharma 1962.

Rasārṇava. See Ray and Kaviratna 1910 and Tripathi 1978.

Rasendracūḍāmaṇi of Somadeva. See Misra 1984.

Ray, Prafulla Candra, and Hariscandra Kaviratna, eds. 1910. *Rasārṇava*. Calcutta: Baptist Mission Press. gretil.sub.uni-goettingen.de. Input by Oliver Hellwig.

Roşu, Arion. 1986. "*Mantra* et *Yantra* dans la médecine et l'alchimie indiennes." *Journal Asiatique* 274 (3–4): 205–268.

Sanderson, Alexis. 1986. "Maṇḍala and Āgamic Identity in the Trika of Kashmir." In *Mantras et diagrammes rituels dans l'Hindouisme*, edited by André Padoux 169–207. Paris: Editions du CNRS.

Sanderson, Alexis. 1988. "Śaivism and the Tantric Traditions." In *The World's Religions*, edited by S. Sutherland et al., 660–704. London: Routledge & Kegan Paul.

Sanderson, Alexis. 2004. "Religion and the State: Śaiva Officiants in the Territory of the Brahmanical Royal Chaplain." *Indo-Iranian Journal* 47: 229–300.

Schwartz, Jason. 2023. "Ending the Śaiva Age. The Rise of the Brāhmaṇa Legalist and the Universalization of Hindu Dharma." PhD diss., University of California–Santa Barbara.

Serbaeva, Olga. 2013. "Can Encounters with *Yoginīs* in the *Jayadrathayāmala* Be Described as Possession?" In *"Yoginī" in South Asia: Interdisciplinary Approaches*, edited by István Keul, 198–212. London: Routledge.

Serbaeva, Olga. 2016. "Varieties of *melaka* in the *Jayadrathayāmala*: Some Reflections on the Terms *haṭha* and *priya*." In *Goddess Traditions in Tantric Hinduism. History, Practice and Doctrine*, edited by Bjarne Wernicke Olesen, 51–73. London: Routledge.

Sharma, Dharmananda, ed. 1962. *Rasaratnasamuccaya*. Varanasi: Motilal Banarsidass. gretil.sub.uni-goettingen.de. Input by Oliver Hellwig.

Shastri, Jagadishlal. 1970. *Kathāsaritsāgaraḥ kaśmīrapradeśavāsinā Śrīrāmabhaṭṭanūdbhavena mahākavi Śrī Somadevabhaṭṭena viracitaḥ*. Delhi: Motilal Banarsidass.

Shastri, Jivaram Kalidas, ed. 1940. *Śrīpārvatīputranityanāthasiddhaviracitaḥ Rasaratnākarāntargataś caturtho* [sic] *Ṛddhikhaṇḍa-Vādikhaṇḍa*. Gondal: Rasashala Aushadhashram.

Shastri, Madhusudhan Kaul, ed. 1923. *The Svacchanda Tantram with Commentary by Kshema Rāja*. Volume 2. Bombay: Nirnay Sagar Press.

Shastri, Madhusudhan Kaul, ed., 1926a. *The Svacchanda Tantram with Commentary by Kshema Rāja*. Volume 3. Bombay: Nirnay Sagar Press.

Shastri, Madhusudhan Kaul, ed. 1926b. *The Netra Tantram with Commentary by Kshemarāja*. Volume 1. Bombay: Tatva Vivechaka Press.

Siddhayogeśvarīmata. See Törszök 1999.

Svacchanda Tantra. See Shastri 1923 and Shastri 1926a.

Ṭhapan, Anita Raina. 1997. *Understanding Gaṇapati: Insights into the Dynamics of a Cult*. New Delhi: Manohar.

Törszök, Judit. 1999. "The Doctrine of Magic Female Spirits—A Critical Edition of Selected Chapters of the *Siddhayogeśvarīmatā* with Annotated Translation and Analysis." PhD diss., Oxford University.

Treloar, Francis E. 1972. "The Use of Mercury in Metal Ritual Objects as a Symbol of Śiva." *Artibus Asiae* 34: 232–240.

Tripathi, Indradeo, ed. 1978. *Rasārṇava nāma Rasatantram (Pārvatīparameśvarasaṃvādātm akam).* Varanasi: Chaukhamba Sanskrit Series Office.

Vasudeva, Somadeva. 2012. "Powers and Identities: Yoga Powers and the Tantric Śaiva Traditions." In *Yoga Powers: Extraordinary Capacities Attained Through Meditation and Concentration*, edited by Knut A. Jacobsen, 265–302. Leiden: Brill.

Verardi, Giovanni. 1994. Homa *and other Fire Rituals in Gandhāra.* Naples: Istituto Universitario Orientale.

White, David Gordon. 1995. "The Ocean of Mercury: An Eleventh-Century Alchemical Text." In *Religions of India in Practice*, edited by Donald S. Lopez, Jr., 281–287. Princeton, NJ: Princeton University Press.

White, David Gordon. 1996. *The Alchemical Body: Siddha Traditions in Medieval India.* Chicago: University of Chicago Press.

White, David Gordon. 2003. *Kiss of the Yoginī: "Tantric Sex" in Its South Asian Contexts.* Chicago: University of Chicago Press.

White, David Gordon. 2009a. *Sinister Yogis.* Chicago: University of Chicago Press.

White, David Gordon. 2009b. "Yoginī." In *Brill Encyclopedia of Hinduism*, vol. 5, 131–135. Leiden: Brill.

White, David Gordon. 2022. *Dæmons Are Forever: Contacts and Exchanges in the Eurasian Pandemonium.* Chicago: University of Chicago Press.

White, David Gordon. 2024. "Three Shades of Tantric Yoga. Chapter 20 of the *Netra Tantra.*" In *The Oxford Handbook of Tantric Studies*, edited by Richard K. Payne and Glen A. Hayes, 369–393. New York: Oxford University Press.

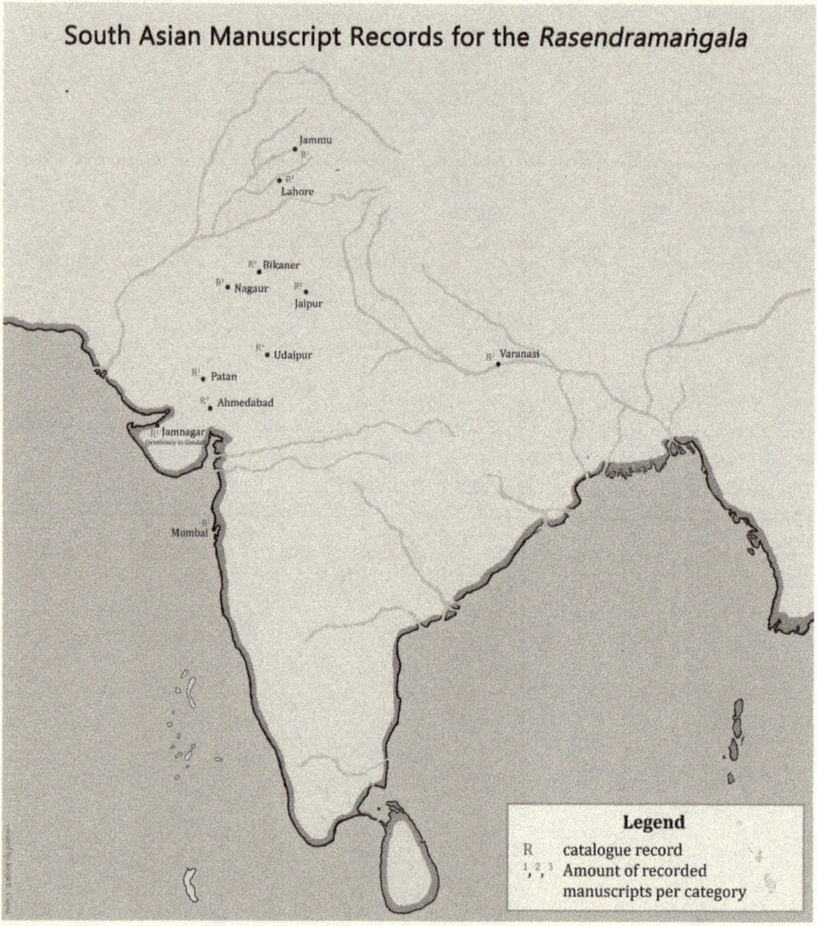

Map 3.1 South Asian manuscript records for the *Rasendramaṅgala*.

3

Nāgārjuna Siddha

The *Kakṣapuṭa* of the *Rasendramaṅgala*

Dominik Wujastyk

The Alchemist Nāgārjuna Siddha

In about 1025, the renowned scholar al-Bīrūnī recorded local information about a renowned alchemist from Gujarat: "A famous representative of this art was Nagarjuna, a native of the fort Daihak, near Somanath. He excelled in it, and composed a book which contains the substance of the whole literature on this subject, and is very rare. He lived nearly a hundred years before our time."[1]

Almost a millennium later, in April 1991, I toured Gujarat with the aim of following up on al-Bīrūnī's remark to see whether I could locate "fort Daihak."[2] My interest in the alchemist Nāgārjuna had begun when I read P. C. Rây's enthusiastic remarks about this author, remarks that I later understood were partly mistaken.[3] Many of the legends concerning Nāgārjuna, as well as the manuscripts of his alchemical work, seemed to come from Gujarat and Rajasthan, so I was hoping that al-Bīrūnī's hints might lead to local discoveries. As I will describe below, it was indeed possible to identify the village mentioned by al-Bīrūnī, where similar legends of Nāgārjuna are still remembered.

Many tales and personalities converge in the figure of Nāgārjuna. These elements appear sometimes together, sometimes separately, in the Sanskrit, Tibetan, Chinese, and medieval Gujarati hagiographic traditions and beyoṇd. Nāgārjuna was depicted by turns as a philosopher-sage, a magician, an alchemist, a Buddhist, a Hindu, and a Jain. He made money, flew through the

[1] Translation by Sachau 1910: vol. 1: 189. On this episode, see also Filliozat (1982).
[2] I was accompanied by Professor Peter Schreiner, whose assistance and language skills were crucial to our success.
[3] Rāy 1902–1909. The work I undertook at that time to disentangle some confusions that Rây inadvertently introduced was published in Wujastyk 1984.

Dominik Wujastyk, *Nāgārjuna Siddha* In: *Indian Alchemy*. Edited by: Dagmar Wujastyk, Oxford University Press.
 DOI: 10.1093/9780197805831.003.0004

sky, cured the sick, created magic gold-producing wells, tricked his teacher, and was finally killed by his son, or by a young prince. At least fifty medical works in Sanskrit and Tibetan are attributed to Nāgārjuna. There have been several attempts to disentangle the many legends and folk tales.[4]

Outside the specific identification of Nāgārjuna as the Madhyamaka philosopher,[5] the name Nāgārjuna is associated with medicine, alchemy, magic, and sorcery, as being the author of "rules for performing various rites for subduing, injuring or killing enemies, and making women complaisant," and "incantations whereby many malevolent designs can be carried out."[6] The medieval literature on alchemy often cited his name specifically in connection with alchemy; one manuscript, for example, calls alchemy "The Nāgārjuna Science."[7]

The earliest surviving testimonies for a specifically magical and medical Nāgārjuna who could make gold and elixirs for extending lifespan include Chinese sources such those composed by Kumārajīva in the early fifth century,[8] Xuanzang in the early seventh century,[9] and the texts and quotations discussed by Walleser (1923) that point to the eighth century.[10] Later Nāgārjuna legends proliferated in Indian and Tibetan literature especially.[11] As Walser remarked, in the context of the famous Buddhist philosopher of the name, "

Despite his fame, it is very difficult to give a straightforward narrative account of Nagarjuna's life, but this is not due to lack of such narratives. The problem is that there are simply too many of them, and they tend to reflect more of the concerns of the communities that preserved them centuries after his life than particularly accurate accounts about Nagarjuna himself."[12] An

[4] Valuable past studies include those by Dās 1882; Cordier 1896; Karambelkar 1952; Yün-Hua 1970; Walleser 1923; Mabbett 1998; Corless 1995. The most comprehensive studies to date are those by Walser 2005; Granoff 1988; White 1996; and especially Meulenbeld 1999–2002, vol. IA: 363–368, 347–363, who also devoted a section to the *Rasendramaṅgala* based on the Paris transcript of a manuscript in the Anup Library, Bikaner, that I was able to supply (Meulenbeld 1999–2002, v. IIA: 714–718).

[5] Surveys include Ruegg 1981: Lindtner 1982; Westerhoff 2022.

[6] Mitra 1880: 574, 588.

[7] "*śrī nāgārjjunī vidyā likhyate,*" MS Pune BORI 942 of 1891–1895 f. 1v, supporting the *Rasaratnākara* (H. D. Sharma 1939: 260, #209).

[8] Walleser 1923: 443ff.

[9] Beal 1884: vol. 2: 214ff.

[10] Walleser 1923: 437ff. Walleser 1923: 440f. raised questions about the dating and origin of the *Mañjuśriyamūlatantra.* Sanderson (2009: 129n300) cited Matsunaga (1985) to argue that the text is datable to the late eighth century.

[11] See further Walleser 1923; Mabbett 1998; Walser 2005. See, for example, Abhayadattaśrī's eleventh- or twelfth-century accounts in *The Legends of the Eighty-four Mahāsiddhas* (Robinson 1979; chs. 16, 18, Donald S. Lopez 2019: chs. 16, 18), or the hagiography by Bu-ston (1290–1364) (Obermiller 1931–1932: vol. 2: 122–130) that was a source for Tāranātha (fl. 1608) (Chimpa, Chattopadhyaya, and Chattopadhyaya 1980: ch. 15).

[12] Walser 2014: 40.

attractively non-essentialist approach to the problem is that of Ollett (2018), who refrains from chasing an elusive historical reality in favor of refocusing the historiography of the Nāgārjuna cycle on the exploration and expression of the essential tension between secular and religious power. This polarity is exemplified in its most pure form in the *Epistle to a Friend* (*Suhṛllekha*), a letter purportedly written by the sage Nāgārjuna to a king.[13]

In addition to the many Buddhist and Hindu associations, Nāgārjuna's name is widely known within Jain circles. The Jain scholar Hemacandra Sūri (fl. ca. 1088–1173), in his commentary on his own *Yogaśāstra*, said that the Jain Āgamas were recorded in writing by the scholars called Skandila Sūri and Nāgārjuna Sūri at a council that was held in Mathurā and is datable to about 770–783.[14] Other sources place the canonical redactor Nāgārjuna (Prakrit: Ṇāgajjuṇa) at Vallabhīpura; this person was the pupil of Himavanta and teacher of Bhūyadiṇṇa.[15] These Jain sources and story-cycles may bear directly on the legends preserved in Gujarat and recorded by al-Bīrūnī.[16]

While in Jamnagar in 1991, I was invited to talk to the postgraduate research seminar of Gujarat Ayurveda University, and subsequently Professor V. J. Thakur (former Vice-Chancellor and dean of the Post Graduate Department) recollected that a colleague of his, Professor Dollarrai R. Makar (Vice-Chancellor of Saurāṣṭra University, Rājkot), had mentioned, forty years earlier, that there might be a connection between Nāgārjuna and Ḍhāṅk. On hearing this name spoken by Professor Thankur, I immediately thought of Al Bīrūnī's "Daihak"; the pronunciation of the vowel in "Ḍhāṅk" was sharp, and when I asked Professor Thakur to clarify his remark, he wrote down for me the name "Dhanka." I knew from Granoff (1988) that the religious biographies of Nāgārjuna included several references to a peak of mount Śatrunjaya called Dhank, but I had not previously connected the name with al Bīrūnī's account or with the village of that name.

On Tuesday 9th of April, Professor Peter Schreiner and I drove to Ḍhāṇk, in Upletā Tāluk, Gujarat, about forty kilometers northwest of Junagaḍh as the crow flies.

[13] See the study by Ye-shes-rdo-rje 2005 and others since Wenzel 1886. A Sanskrit manuscript of the work has recently been reported by Kano 2020.

[14] Kapadia 1941: 61–62. On the context of such assertions, see Wiles 2006.

[15] Mehta and Chandra 1970–1972: vol. 1: 316, drawing on the *Daśavaikālikacūrṇi* 1933; *Ācārāṅgacūrṇi* 1941; *Ācārāṅgavṛtti* by Śīlāṅka 1941; *Uttarādhyāyanacūrṇi* 1933; *Uttarādhyāyanavṛtti* by Śāntisūri 1916.

[16] On Jain and Gujarati Nāgārjuna-cycle tales, see Granoff 1988.

Shortly after entering the village, we were surrounded by helpful villagers, and we asked if there was a museum or any ruins in Ḍhāṅk. We were told there were not. We decided to drive up to a temple visible on a nearby hill to the northeast, outside the village. It was a fairly modern Śiva temple. While there, we asked the clerk of the local school, Bhagavanjī Bachu Bhai Sutariya, who had accompanied us, whether there were any old caves at Ḍhāṅk. He said there were, and led us further over the hill about half a kilometer beyond the temple. There we found a sudden cleft in the ground, a small ravine, with old carved steps in the side leading down into the cleft. On the west side of the ravine, which runs approximately north-south, were a series of obviously ancient monastic caves. Inside one was an image of a Jina, carved out of the cave wall. Inside another was a hole about half a meter across, above head level, disappearing into darkness.

Later study revealed that these caves were first documented in the nineteenth century by James Burgess, who thought they were Buddhist, and subsequently by the great archaeologist Sankalia, who called them the earliest such caves in Gujarat.[17]

At that point, Professor Schreiner asked Mr Sutariya if there were any legends about Nāgārjuna. To our astonishment, Mr Sutariya said yes indeed, that Nāgārjuna was a famous alchemist and had had an alchemical premises (*rasaśālā*) in Ḍhāṅk. Mr Sutariya told the following stories, sometimes after consultation with other villagers.[18]

> There was a sage called Ḍhuṇḍhurinātha who lived on a nearby hill. In those days, Ḍhāṅk was called Preya Pātan. Ḍhuṇḍhurinātha had a pupil called Siddhanātha, and Ḍhuṇḍhurinātha had a famous alchemical premises (*rasaśālā*). Siddhanātha was sent begging in Ḍhāṅk, for alms (*bhikṣā*), but got nothing. As a punishment for the ungenerous town, Ḍhuṇḍhurinātha cursed Ḍhāṅk, that it should be turned to ashes. Nāgārjuna was the yuvarāja, prince, of the king. He was of the Jaiṭhvā caste.

[17] Burgess 1876: 2: 150–152 and Sankalia 1938. Sankalia at first dated them to the fourth century, but later revised this to the first or second century.

[18] Mr Pravinbhai Ukani, a resident of Ḍhāṅk and landowner, interpreted into English, although he did not know the stories himself, and was most helpful and hospitable. Mr Sutariya later said that these tales of Nāgārjuna were written in a Gujarātī book called *Saurāṣṭra Rāsdhār*. Later, he produced the book, and it was Śambhuprasād Haraprasāda Desai's *Saurāṣṭrano itihāsa* [History of Saurāṣṭra], published by Soraôh Śikṣan aṇe Sanskrit Sangha, Junagaḍh, 1968.

> In those days, Ḍhāṅk was big. It was called Preya Pātan. Today it is underground. Whenever foundations are laid for new houses, old walls are found in the ground. Underground there are thousand-year-old statues (*mūrti*) of Jains and Bauddhas, in black stone. When people dig them up, they keep them in their homes. There are old wells too.
>
> [I asked for the story to be repeated:]
>
> Ḍhuṇḍhurinātha had a pupil called Siddhanātha. He was begging in the town for alms. Ḍhuṇḍhurinātha was meditating. At that time, this city was mainly Jains and Bauddhas. Ḍhuṇḍhurinātha was a follower of Brahmanism (*sanātana*). So the people wouldn't give alms. Ḍhuṇḍhurinātha became angry, and cursed the city so that Preya Pātan would be turned to dust, or would collapse. Even today, people find pots of dust in the ground, and this dust is what used to be money.
>
> Nāgārjuna's mother was Queen Mother (*Rājamātā*) of Preyas Pātan. The Rājamātā was crying. She went to Siddhanātha and complained about Ḍhuṇḍhurinātha's curse. Siddhanātha said "it is done." But he offered to help create a new city, and that is Ḍhāṅk. In Gujarātī, "to go to *ḍhāṅk*" still means "to go bad."
>
> Nāgārjuna said, "I want to make this city large and comfortable." He made gold and he wanted to make the walls themselves of gold.
>
> Nāgārjuna and his mother went to Siddhanātha and Ḍhunḍhurinātha, who gave them an alchemical (*rasa*) formula to make gold and so on.
>
> Nāgārjuna's son wanted to kill him. The son attacked him with a sword, but couldn't kill him. So Nāgārjuna put some metal from his mouth onto the blade, and the son was then able to cut off his head. *That head can be seen in Ḍhāṅk today.*
>
> Siddhanātha made a tunnel from Ḍhāṅk to Junagāḍh. It is the one that begins in one of the caves we saw earlier.

We were shown another small hill, beside the dirt track leading to the Śiva temple hill, and told that this was the hill on which Ḍhunḍhurinātha had lived and meditated. We were also told that the villagers had a strong belief that the hill probably covered remains from his life. Light-colored earth on the hillside was identified as being ash from his fireplace. Apparently the villagers had asked the Archaeological Survey of India to investigate the mound, but the initiative had been turned down.

After the above stories were told, Professor Schreiner and I were taken to the home of the lady in Ḍhaṅk who had Nāgārjuna's head in her home. We were shown to a whitewashed room where there were two heads in current

Figure 3.1 Nāgārjuna and Baroṭ, private home, Ḍhāṅk, 1991. Photo © Dominik Wujastyk.

worship (Figure 3.1). They were both painted orange/red, and had eyes attached in the Gujarātī fashion. The head of Nāgārjuna "Jaiṭhva" was on the left, and appeared to be an old piece of Jain sculpture, worn smooth by time and worship.[19] There were traces of a headdress, and features on the face, which was round in the manner of some Jain sculptures. We were told that the torso that belonged to this head was in Palitāna.

The second head was not immediately recognizable as a sculpture, and may have been simply a stone. It was said to be the head of a Baroṭ, who was the person who wrote the life story of Nāgārjuna.[20]

Once these connections had been made, further records of Nāgārjuna the alchemist in Gujarat came into focus, dating from the eleventh to the nineteenth centuries.[21] The eleventh-century *Kathāsaritsāgara* devoted a chapter to Nāgārjuna's exploits, which include making the "water of immortality"

[19] Jeṭhvas are a Kathiavad community with ancient roots.

[20] On the Baroṭ community, traditional historians, see Forbes 1856: vol. 2: 262–267; Shah, Shroff, and Srinivas 1958; and Thompson 1992. Outside this house, on the wall facing the front door, was another sculpture, inset in the wall. A face was just discernible and it was connected with Nāgārjuna in some way.

[21] See Granoff 1988.

that "prevents mortals from dying."[22] An elaborate tale of Nāgārjuna that has much in common with the Tibetan tales occurs in Merutuṅga's *Prabandhacintāmaṇi*, completed in 1304, including explicit reference to Ḍhāṅka.[23] When European observers began collecting tales in Kathiawad, the Nāgārjuna stories were also recorded. Forbes, recording bardic historical narratives in western India in the mid-nineteenth century, recounted the narratives of "Dhoondule Mall the sage" and the fall of Wullubheepoor (Vallabhi), which is a transferred variant of the Ḍhuṇḍhurinātha/Ḍhāṅk account, complete with the idea that modern diggers find the walls and bricks of the old city.[24] Vallabhī and Ḍhāṅk are linked elsewhere in the *Rās Mālā*, in tales that Forbes dated, in some cases, to the mid-second century.[25] Wilberforce-Bell, collecting narratives at the beginning of the twentieth century, thought that Nagarjana (*sic*) Jethwa "held sway" in Dhāṅk in the first century.[26]

It seems beyond doubt that in Gujarat in 1025, al-Bīrūnī was in receipt of the same type of local Nāgārjuna legends connecting the alchemist-sage with the town of Ḍhāṅk that have been recorded over several centuries in Sanskrit and Gujarati literature, in the reports of early European visitors and that the kind villagers of Ḍhāṅk shared with Professor Schreiner and me almost a millennium after al-Bīrūnī.

The *Rasendramaṅgala*

Can we connect these ancient and modern legends from western India about the alchemist Nāgārjuna with the *Rasendramaṅgala* text ascribed to Nāgārjuna Siddha? The first and most tangible argument can be made from the locations of the surviving manuscripts. They come almost without exception from Rajasthan and Gujarat.[27] This fact, coupled with the almost complete absence of manuscripts from anywhere else, strongly suggests that the work originated in this part of the world. The text itself does not mention Dhāṅk, but locates part of its narrative at the mountain of Śrīśailam. This is

[22] Durgāprasād and Parab 1889: 216–217; trans. Tawney and Penzer 1924–1928; vol. 3: 252–256.
[23] Muni Jinavijaya 1933: 1: ch. 5; trans. Tawney 1901: 28, 194–196.
[24] Forbes 1856: 13–15.
[25] Ibid.: 21, 295, 296.
[26] Wilberforce-Bell 1916: 49–52.
[27] Locations include Ahmedabad, Bikaner, Mumbai, Jaipur, Udaipur, Gondal, Patan, Koba, Jammu, Bikaner, and Nagaur (Wujastyk 2023). See the map above, to which two manuscripts, held in Koba, can now be added.

one of the features of the *Rasendramaṅgala* that conform to the Nāgārjuna legend in its form that preceded its adaptation by Jain authors.[28]

The last part of the *Rasendramaṅgala*, presented below, is framed as a dialogue between Nāgārjuna, a Wood Nymph (*Vaṭayakṣiṇī*), and a king called Sālivāhana.[29] In his discussion with the Wood Nymph, Nāgārjuna is granted a wish. He asks for the means to perform wonders and for the ascetic discipline of mercury (*rasasādhana*) to conquer death and poverty. He then says that he wishes to make golden the palace apartment of Pārvata, which has thickets and wooded mountains. This dialogue takes place on top of Mount Śrīśaila, where Nāgārjuna teaches alchemical doctrines to an assembly of fifty listeners including Ratnaghoṣa and Śūrasena.[30] At the end, King Śālivāhana declares that he is ready to sacrifice all his possessions to the alchemical art, and even his son and Madasundarī. Other personalities about whom stories are related include Māṇḍavya (usually a mere name in authorities lists), Vasiṣṭha, Ratnaghoṣa, and Nāgabuddhi.[31]

The work is throughout redolent of a Buddhist milieu, and nowhere is this more striking than in the passage where Prajñāpāramitā appears in a dream.[32] However, other religious elements are also present, and at its end, the dialogue introduces Bhairava.

All these features remind us strongly not only of the Jain Nāgārjuna legends of western India, but also of the broader cycle of alchemical Nāgārjuna legends from known India and Tibet, including especially the relationship between Nāgārjuna and a king, the production of wealth and longevity, and the environment of caves, rocks, and holy mountains, as well as a junior disciple, and a final death by beheading.

Granoff has suggested that the reputation of a Nāgārjuna associated with a Jain alchemical work was a factor that led Jain authors to absorb Nāgārjuna into their *imaginaire*. If this is so, we may be seeing evidence here for a period before 1196 , when the *Kumārapālapratibodha* provided the earliest datable version of Nāgārjuna's alchemical adventures in Jain Prakrit literature from Gujarat, including his conversion from Buddhism to Jainism.[33]

[28] For this adaptation, see Granoff 1988: 47–49 et passim.

[29] While "*vaṭayakṣiṇī*" more specifically means "Banyan Nymph," the translation "Wood Nymph" is more evocative for the English reader. Dwivedi (2022) has drawn attention to the widespread Sanskrit trope of tree nymphs as tutelary spirits.

[30] On the importance of Śrīśaila as a source of alchemical traditions, see Roşu 1992. There may be connections to be made—historical or mythic—between "Dhāṅk" and "Dhānyakaṭaka."

[31] Wujastyk 2023, passim.

[32] See the summary of chapter 3, below.

[33] Granoff 1988: 54, 60; Munirāja Jinavijaya 1920: 179.

The Contents of the Rasendramaṅgala

The opening verses of the *Rasendramaṅgala* describe it as a work having eight chapters, and sketches their contents as follows:[34]

1. On the cleaning (*śodhana*) of mercury and its subsidiary substances (*uparasa*);
2. on the calcination (*māraṇa*) of diamond (*vajra*) and so on; on the liquefaction (*drāvaṇa*) of all elements;
3. on the binding (*bandhana*) of mercury;
4. on pills (*guṭikā*);
5. on the medical treatment of wind and so on, in all diseases;
6. on perfumes and salves (*añjana*);
7. on poisons;
8. on qualities (*guṇa*).

Unfortunately, all the available manuscripts of the *Rasendramaṅgala*, including the earliest dated one from the seventeenth century, contain only the first four chapters. This is all that has been transmitted to the present day, barring further discoveries.[35]

After a general introduction and the above self-description, chapter 1 of the *Rasendramaṅgala* describes the eight classical magical powers, from becoming tiny to having power over others. Mercury can deliver all these powers, and it can cure all illnesses, remove life-obstacles, give pleasure and freedom and immortality. It can make a man the equal of the god of love. Of what use is a physician who does not know about mercury? Mercury is the most powerful drug of all, and gives greater power to all other drugs.

Mercury may have impurities or other blemishes that cause nasty physical effects. So there are eighteen methods of processing mercury.[36] These processes are described, as are further recipes and purifications involving gold, silver, copper, bronze, lead, and tin.

[34] A technical overview of the work was provided by Meulenbeld 1999–2002, IIA: 714–718.

[35] Wujastyk (2023). It is interesting to note that the missing chapters cover traditional major topics of the Ayurvedic tradition.

[36] The *saṃskāras* listed in 1.22–23 are: 1. *saṃsvedana*, 2. *mardana*, 3. *mūrcchana*, 4. *utthāpana*, 5. *pātana*, 6. *dīpana*, 7. *niyāmana*, 8. *rodhana*, 9. *cāraṇa*, 10. *garbhadruti*, 11. *bāhyadruti*, 12. *jāraṇa*, 13. *grāsapramāṇa*, 14. *rañjana*, 15. *saṃsāraṇa*, 16. *krāmaṇa*, 17. *vedha*, and 18. *yoga*.

Chapter 2 describes further elaborate recipes and procedures, now also including diamond as well as human and animal products such as menstrual fluid, breast milk, and horse urine. The process is described as heating materials at high temperature in a crucible to extract their essences (*sattvapātana*).[37] Further crucible-based processes are described involving honey, clarified butter, castor-oil, cow-dung, and many other substances. These processes may take several weeks to complete. The chapter ends with the week-long processing of pearls into an empowered liquid.

Chapter 3 takes a medical turn, starting with metal and plant recipes that cure humoral disturbances in wind, bile, and phlegm. Some recipes will destroy disordered humors, "like a thunderbolt striking a tree."[38] Unusually for a technical treatise, elaborate literary poetic meters, especially Śārdūlavikrīḍita, are found throughout the *Rasendramaṅgala*, but they are particularly notable in this section. The repeating structure of the verses is to describe the preparation in detail and to end with a short statement of its predicted therapeutic effect. This chapter includes a distinctly Buddhist reference, invoking Prajñāpāramitā: "In a dream, during the night time of Prajñāpāramitā, the wise Śrī Nāgabodhi manifested this pure king of essences called Sharp-mouth."[39]

After an initial section describing medicines that treat humoral imbalances, the author self-consciously turns to a different ordering principle: "Until now I have written about therapy according to humours; from here onwards I shall write about it in order of disease." The author then proceeds with therapies for indigestion, fevers, fluxes, and so on, including several panaceas. In rare reference to the alchemist's patient, the author states that a single dose of a particular gold, sulfur, and mercury ash will destroy all diseases and old age in the Seeker-King (*sādhakendra*) and give him a divine body (*divyadeha*).[40] There is a long section on recipes using tourmaline (*vaikrānta*).[41] Among further therapies for illness, old age, and poverty

[37] Hellwig 2009a: 59, et passim, has shown that essence-extraction (*sattvapātana*) is a tell for the earliest Sanskrit alchemical works, the process becoming almost obsolete in late Indian alchemy and being displaced by an increase in calcination operations (*māraṇa*). Using this criterion, the *Rasendramaṅgala* would be grouped chronologically with the earliest works such as the *Rasahṛdayatantra*, the *Rasārṇava*, and the *Rasādhyāya*.

[38] *Rasendramaṅgala* online edition (hence RM*ed*) 3.8 (Śarmā 2003: 52).

[39] Provisional edition: *prajñāpāramitāniśīthasamaye svapne prasādīkṛtaṃ / nāmnā tīkṣṇamukhaṃ raseṃdramamalaṃ śrīnāgabodhairbudhaiḥ//*. Cf. the statement, "there [at Śrīpārvata], the deity manifests during a dream and grants a boon," (*tatra svapne devatā pratyakṣībhūya varaṃ dadāti*), *Rasaratnākara, Rasāyanakhaṇḍa, Śrīśailasādhana*, 185; Ācārya 1939: 78).

[40] RM*ed* 3.35 (Śarmā 2003: 63).

[41] The identification of *vaikrānta* is uncertain.

are found recipes for creating gold ;[42] for turning an old man into a sixteen-year-old ;[43] for restoring hair, nails, and teeth; and for generating divine eyesight (*divyadṛṣṭi*) and a lifetime of a century.[44] A number of recipes treat skin diseases including leprosy and pachydermia (*gajacarma*), and others treat more quotidian ailments such as cough, wheeze, hiccups, heartburn, painful urination, and also epilepsy and insanity. A number of eye-salves contain copper among other ingredients or are to be prepared in a copper vessel.[45]

A striking passage occurring in this chapter in some of the witnesses is the description of an eye-ointment that was widely quoted as the "Nāgārjuna's Wick" recipe by medieval Bengali authors such as Vṛnda (fl. 800–950) and Cakrapāṇidatta (fl. 1075).[46] This recipe became famous and was cited by many later authors.[47] The Nāgārjuna's Wick verses occur in some manuscripts of the *Rasendramaṅgala*.[48] The recipe consists of fourteen ingredients, including copper and copper sulfate, that are powdered together and pressed into wicks with rain water. The verses curiously assert that this recipe was "inscribed by Nāgārjuna on a pillar in the city of Pāṭaliputra" (today's Patna).[49] The chapter finishes with a small group of recipes for treating ear, nose, and throat ailments, as well as disease of the vagina and demons that attack children.

Chapter 3, with its strong medical orientation, covers topics that were announced at the start of the *Rasendramaṅgala* as being the content of chapters 5 and 7.

With chapter 4 we move into a quite different atmosphere. No longer are the recipes therapeutic: now we are in the realm of chemistry and magic.

[42] RM*ed* 3.44–47 (Śarmā 2003: 66).

[43] RM*ed* 3.56 (Śarmā 2003: 69).

[44] Rm*ed* 3.78–79 (Śarmā 2003: 71).

[45] Compare with the effective eye-salve including copper from the Anglo-Saxon *Bald's Leechbook* researched by Harrison et al. 2015 and discussed by Wilson 2015.

[46] *Siddhayoga* 61.148–152 (Pādhye 1894: 470); *Cakradatta* ch. 59, verses 125–126 (P. V. Sharma 1994: 499). See Meulenbeld 1999–2002, IA: 364, IIB: 477.

[47] E.g., in Cāmuṇḍa's *Rasasaṅketakalikā* 5.25–27 (ca. 1490), *Rasaoyogasāgara* takārādi #406. See the further discussion by Filliozat 1979: v ff.

[48] For example in MS Gondal BP 34 f. 18r, and apparently in the unidentified Bikaner MS used by H. Ś. Sharma 1999: 111, vol. 3.201–202. This distribution suggests that the passage was added to late manuscripts by a later hand, and was probably not part of the original text.

[49] Many ancient pillars have been excavated at Patna (Kumrahar site) since the time of Spooner's discovery of an eighty-pillared hall in the early twentieth century. It is tempting to connect this Nāgārjuna legend with the archaeological discovery of a Buddhist hospital (*ārogyavihāra*) and associated inscription in Kumrahar, dated to the period ca. 300–450 (Altekar and Mishra 1959: 41, 53), but there is no explicit link as yet.

A long opening section lays out a complex ten-step program for using a compound based mainly on diamond, mercury, and gold and processing it into a solid form.[50] At each of the ten steps one may take an increasingly mercury-rich and increasingly powerful compound. By holding it in one's mouth for periods from one month to nine months, one may increase one's lifespan from a yuga through the lifetimes of the gods and finally become lordly, all-pervading, a creator, destroyer, experiencer, one who bestows both curses and favors. The next recipe makes a man victorious in battle, loved by all, and attractive to women. The author refers here to having learned the recipe from "the sages (*sūri*)," using a term that most often refers to Jain scholars. Recipes appear for the first time that include animal parts such as the hearts and eyeballs of several kinds of birds. Crushed and worn on the throat, this recipe allows the alchemist to travel hundreds of miles unseen. The theme of making aphrodisiacs and compounds that will attract women continues throughout this chapter. Another pill, held in the mouth for a whole year, removes all flaws and leads to a lifetime as long as a yuga. One recipe calls for a mixture of gems and menstrual blood to be made into a wax-covered ball that will give success in battle, with the royal family, in gambling, and in love. The theme of delivering a divine body (*divyadeha*) recurs repeatedly, as does relief from old age and illness. A common feature of the chemical operations is the assertion of a parallelism between the chemical transformations and similar enhancements of the human experimenter. The dictum, "as in the metal, so in the body," is found more than once in the *Rasendramaṅgala*, and it is a saying known from other alchemical works.[51] Processes of liquefying and binding (*drutibandhana*) mercury create an essence that perfects the body and metal alike.[52] In addition to very worldly goals, this chapter also offers the fulfilment of the canonical four goals of righteousness, wealth, pleasure. and liberation.[53] Verse 4.115 is structurally interesting because it contains a cross-reference to processes described earlier, in chapter 2, demonstrating some internal coherence at the level of the work.[54] After ending a section on liquefaction and amalgamation, the author turns to numerous recipes with

[50] This solid mercury form (*khoṭa*) is described in the *Rasajalanidhi* as one of the twenty-five ways in which mercury can be solidified (*baddha*) and is achieved by being reduced by condensation (*dhmāta*) and diminution (*kṣaya*) (Mookerji 1926–1938: vol. 1: 314).

[51] *Yathā lohe tathā dehe* . . . ; RM*ed* 4.72ab (Śarmā 2003: 134, et passim). Found, for example, in the *Rasārṇava* (14.16) and the *Ānandakanda* (1.23.615, 639).

[52] RM*ed* 4.155 (Śarmā 2003: 152).

[53] RM*ed* 4.102 (Śarmā 2003: 141).

[54] RM*ed* 4.115 (Śarmā 2003: 144).

direct sexual applications, mainly aimed at controlling women. Some aim to remove a woman's craziness and vanity,[55] others to give her pleasure at different stages of intercourse,[56] while others provide vaginal lubrication. A man with one recipe smeared on his hand can cause a woman to relax just by grasping her hand.[57] Other recipes enlarge both a man's penis and a woman's breasts.[58] The topic now abruptly returns to several recipes for multiplying quantities of gold.[59]

The chapter next continues with a subsection entitled the *Kakṣapuṭa*.[60] The sixty verses that follow begin with a verse that in some of the older manuscripts invokes "all the Buddhas" (*praṇipatya sarvabuddhān*) and in others invokes the great god Mahādeva instead.[61] The verses promise that the *Kakṣapuṭa* will deliver extraordinary powers for the good of all people. The person of Nāgabuddhi is mentioned twice, once in connection with transmutations and again as one to whom an alchemical formula is explained. In the latter passage, a recommendation is made that one should show the Buddhist virtues of benevolence, compassion, joyousness, and equanimity toward all beings.[62] The setting and tone now change mid-chapter, and we are transported to the peak of Mount Śrīśaila, where a dialogue takes place and Nāgārjuna is taught powerful alchemical techniques by a Wood Nymph, so that he can create wealth for King Śālivāhana.[63] This is the passage translated below.

[55] RMed 4.157 (Śarmā 2003: 154).
[56] RMed 4.166 (Śarmā 2003: 156).
[57] RMed 4.182 (Śarmā 2003: 162).
[58] RMed 4.187 (Śarmā 2003: 163).
[59] The edition of Śarmā (2003) ends here, but almost all manuscripts continue with many more verses.
[60] I am not certain of the meaning of this title in this context, and it is not explained by our authors. It could signify "hidden pit," "a box with divisions," or even "armpit" (Goudriaan and Gupta 1981: 117), or a number of other interpretations. This text is not related to the *Kakṣapuṭa* of twenty chapters that begins *yaḥ śāntaḥ paramālayaḥ* (Wujastyk 1984: 75 et passim; Goudriaan and Gupta 1981: 117–118; Bhaṭṭācārya and Bhaṭṭācārya 1915: 264–390; Khandelavala 1996).
[61] The former reading is in approximate Āryā meter, the latter in ordinary śloka. We may be seeing here the process of turning a Buddhist text into a Brāhmaṇa one.
[62] *Karttavyaṃ maitrīkaruṇāmuditopekṣā sarvasatveṣu*. On the history of the Brahmavihāras see, e.g., Maithrimurthi 1999.
[63] Secondary literature on Mount Śrīśaila is provided by Meulenbeld 1999–2002, vol. IIB: 668n130. The fourth part of the *Rasaratnākara*, the *Rasāyanakhaṇḍa* (fifteenth century), has a chapter dedicated to special alchemical activities to be performed at the various gates and locations at Mount Śrīśaila (Ācārya and Paṇaśīkara 1939: *upadeśa* 8: 65–78; see notes by Meulenbeld 1999–2002, vol. IIA: 658, 663–664, and the chapter by Cantú in the present volume).

Notes on the Text and Translation

This passage has not been edited, printed, or translated before. The translation that follows is based on my own collation and preliminary editing of the dozen surviving accessible manuscripts of the work.[64] The text has not been well served by the manuscript scribes. Many passages remain obscure to me, and that lack of clarity is reproduced in the translation below. As was noted by Newman (2018: 11), Newton's alchemy "cannot be deciphered by linguistic means alone; it requires a knowledge of materials, technologies, and tacit practices as well as underlying theories submerged beneath the written word." It is much the same with the *Rasendramaṅgala*. Hellwig's rich dictionary (2009b) and the practical discoveries and experiments filmed and documented by Dagmar Wujastyk and the members of the AyurYog project (Ayuryog.org) have been of enormous help in interpreting the text. But much remains obscure to me, hidden by layers of non-standard Sanskrit, scribal misreadings, and tacit practices and theories. To my chagrin, I have not been able to make sense of every verse. The thickets of variant manuscript readings, verbless sentences, strange lexemes, mis-matched concord, and so on often baffled me. I have had to leave out many verses for this reason, while my interpretation of several others—marked with daggers—is very doubtful.[65] Due to this and the constraints of time, I have offered only a summary of the last part of this chapter. I commend this difficult text to future editors and translators.

[64] The manuscript transcriptions are available at Github and Zenodo. See https://wujastyk.github.io/Rasendramangala (https://doi.org/10.5281/zenodo.6484115). The edition can be consulted at https://saktumiva.org/wiki/wujastyk/rasendramangala/chapter4-srisaila/provisional_edition.

[65] Omitted verses include several from the wisdom of Markaṇḍeya (verses 71ff.) that are parallel to verses in the *Rasārṇava* and the *Ānandakanda*, where they are equally obscure. For example, RĀ: 12.364–365 (Ray and Kaviratna 1985) = ĀK 1.23.565–567 (Radhakrishna 1952). I have also omitted most verses after no. 117 for lack of time.

Rasendramaṅgala, Chapter 4, the Kakṣapuṭa Section

On Mount Śrīśaila

1. The great Siddha Nāgārjuna, endowed with all the qualities of wealth, a benefactor of all beings, stood on Mount Śrīśaila.
2. Upon request, he quickly grants[66] anyone whatever they ask, after seeing renunciation and enjoyment by the grace of mercury.
3. He is the great transmuter of all essences, †O you who have an army of bards, for it is like yours†.[67] Ratnaghoṣa Prabhākara was the chief among them.[68]
4. Holding his hands respectfully cupped, standing in front of Nāgārjuna, he asks about the mercurial procedures. "Give me the gift of knowledge."

Nāgārjuna said:

5–6. Bravo, bravo, you of great wisdom! I am delighted, you who are attached to devotion. Have no doubt: I will explain what you have asked about, the removal of wrinkles and gray hair, the cheating of time. As it proceeds in metal, so it is in the body; there is no doubt about this.[69]

7. Mercury can transform a million times, a thousand, myriad, a hundred-thousand times. I will explain that accomplishment to you in the proper way.
8. In order to awaken beings, a Wood-nymph was sought. And I performed twelve years of great suffering.[70]
9. At the end of that time I heard a heavenly voice belonging to invisible divine beings. Then I begged the unseen lady, "now become visible."

The Wood-nymph said:

[66] *Dadhate*, √*dhā* as a 1. conj.

[67] Or, "a man of Mathurā," reading *śūrasenas* against the MSS. The text of the first part of this verse is unclear.

[68] Ratnaghoṣa is mentioned in a list of alchemical authorities in the *Rasaratnasamuccaya*, verse 6.51 (Vidyābhūṣaṇa and Vidyāratna 1927: 179), immediately following Nāgārjuna.

[69] A verse that occurs in several works, including the *Rasārṇava*, the *Sarvadarśanasaṅgraha*, and the *Ānandakanda*.

[70] Possibly related to the twelve-year famine in northern India reported in epigraphical records and in Jain literature that took place during the time of Skaṇḍila Śūri (Virji 1952: 179).

10. Bravo, bravo, a great bravo! Because of your devotion, which is your greatness, I shall give you everything that you have yearned for regarding what is to be done, my dear.

Nāgārjuna said:

11. O Goddess, if you are pleased with me, you who have been tormented for twelve years, † Give me the truth about yourself. After that, I shall be free.†[71]

[The Wood-nymph said:]

12. True, true, and again true! The word was spoken three times. Whatever you ask for, O Siddha, I will give you all that.

Nāgārjuna said:

13. O Goddess, you are fond of devotion; if my devotion pleases you, grant me the binding of mercury that is so hard to get in the three worlds.[72]
14. I shall perform a great wonder, by any means whatsoever, as well as the achievement of mercury that overcomes death and poverty.
15. †Out of a desire for the welfare of the world I shall turn into gold the mountains having houses and palaces, with groves, forests, and peaks.[73]
16. I shall give food, clothes, and betel nut to my own friends. I shall make fame for myself in this world.
17. O elegant Goddess, please describe the method.

The Wood-nymph said:

18. I shall give you whatever you ask for, O Siddha. Carry out the procedure properly; I shall be standing at your side.

[71] The verb *dhuvāmy* (cl. 6) is doubtful here.

[72] The "binding" of mercury refers to a group of preparatory processes that render mercury safe and ready for digestion by other substances. See Hellwig (2009b: 64ff.).

[73] There are grammatical problems with this verse, and the translation is tentative.

19. A person who is replete with all good signs, who nourishes an auspicious person, will have your success by a touch and a glance.
20. After hearing his statement I asked Śālivāhana, "by using your power I shall make the great essence."[74]

Śālivahana said:

21. I entrust my storehouse of gold and jewels to myself. † Beautiful girl . . . † Let the command be given, O King.

[The Wood-nymph said:]

22. Bravo, bravo, O Wise One. You are a guardian of my commands. Have no doubt: I will bring about success, because of your truth, O Seeker.
23. Moreover, I shall explain another perfected quicksilver that was made by Māṇḍavya, using essences and subsidiary essences.[75]
24. For the purpose of ritual, transmuted copper and iron and lead was made into gold. That quicksilver method was learned by Vaśiṣṭha, who was at his side.
25. I heard the treatise of Vasiṣṭha and Māṇḍavya at my teacher's side. So I shall correctly explain that process.
26. The companions are beautiful and wise, energetic and firm in their commitments. They are of good family, sinless, full of righteousness, and self-controlled.

27–28. One should make a collection of all these items: a little oven, a bent-tube apparatus (*vakranāla*),[76] cow-dung coals, kindling, a pipe, sheets of metal, herbs, sour gruel, black salt, varied camphors, and various crucibles too. One should take up that work.

29–31. There is a gateway to a courtyard, a pot of garlands together with water. One should install the great sage on a throne standing in the middle of the courtyard, with limbs smeared with cool sandal paste, wrapped in white clothing, a powder that is the same as five

[74] There is some confusion here about the speakers.
[75] Māṇḍavya is mentioned in a list of alchemical authorities in the *Rasaratnasamuccaya*, verse 6.52 (Vidyābhūṣaṇa and Vidyāratna 1927: 179).
[76] Hellwig 2009b: 431.

jewels together with a lamp and something undamaged, with the incenses of sandalwood and agarwood (*aguru*), and various kinds of food offerings.

32. Next, Great Wisdom should be prepared and made to eat food for a long time. Young girls should also be worshiped, but the man should ask permission.
33. By the grace of your command I shall succeed with the great essence. Now, at the beginning, one should take purified quicksilver that is without blemish and firm.
34. Then one should undertake an appropriate procedure on a substance that is a longevity tonic (*rasāyana*). Then that quicksilver (*sūtam*) is cleansed well, is extracted, and has discharged all faults.
35. Quicksilver that has been digested eight times by mixing with the essence of calamine (*rasakasatva*) should be stirred, and then the man should cause it to be digested with an equal amount of black mica.[77]
36. One should offer the same to sulfur. One should burn it up with fire and soma. One should cause wrought iron (*tīkṣṇa*), copper, and good lead to be returned (*samāvṛtta*).
37. Placed in the middle of a steam apparatus, it should be aged with an effort. Furthermore, another should be prepared so that it becomes gold.
38. She is famously called Gokarṇī; the second is Kṛṣṇamañjarī. These are to be grasped at any opportunity: there is no perceptible difference.
39. At whatever location she arises, on the earth or on a mountain, she is pointed out at that location, which is how she can be fruitful.
40. Therefore, one should make every effort with the visible mantra.
41. The mantra.

Oṃ, a bow to you, who is equal to death, who magnifies strength and heroism. Give me strength and life. Cast sins far away from me.

42. *I will dig this with the same five-fingered hand that Brahma digs with, that Hari digs with.*
43. The mantra.

[77] On the possible identities of *rasaka-satva* see Hellwig (2009b: 320–323). "Eight times" *aṣṭaguṇe* is grammatically irregular.

† *In the eightfold one that has flown up and fallen, stay here, stay here, O Queen of the Gods. After helping me succeed with the work you will then go to heaven.*

44. One should make it consecrated seven times with this mantra. †After heating it by contact with the sun, one should dry it in the shade.
45. As long as she does not become ash by means of sour milk with Mahua,[78] the goddess who destroys death and poverty will have all her parts.[79]
46. The herb should be amalgamated in mercury and gold that is located in the middle of the steam appliance. Alternatively, the wise person can do so in combination with *viḍa*-salt.
47. Fire above and water below with the mercury held in the middle. Gold appears in the mercury after applying the fire again and again.
48. By means of this stepwise procedure (*kramayoga*) carried out for four days, it will expand (*sphoṭaya*) in the middle of the water. After washing it, make it pure.
49. One should grind together datura (*unmatta*), cucumber leaves (*munipatra*), turmeric (*rajanī*), and Deccan hemp (*kācamācikā*)[80] in equal measures, with sour rice gruel (*āranāla*).[81]
50. By means of this stepwise procedure (*kramayoga*) carried out for seven days, † then produce double it with the juice of the heavenly herb.
51. The wise person first pulverizes the mercury that looks like a hen's egg. He should then place the †four that are sap (*niryāsa*) of the Brahma flower (*brahmapuṣpa*) into the baking pit (*puṭa*). Thus there is the sap made from Indian madder (*mañjisthā*). Do it like this.
52. One should have the four put into the baking pit (*puṭa*) with the sap of five-leaved chaste tree (*śephālika*). Do exactly the same with the mercury of Kashmir.
53. One should have the one put into the baking pit (*puṭa*) with the juice of a citron (*mātuliṅga*). From that one may know that it has succeeded. One should give the transmutation (*vedha*) of the copper (*śulva*).

[78] "*Madhukāṣṭha*" is a rare synonym for the Madhūka tree, or *Madhuca longifolia* (J. konig) J. F. Macbr. It has sweet flowers.

[79] Or, "that goddess Sakalā who destroys death and poverty will appear."

[80] Normally, *kākamācikā*.

[81] Similar to *Rasaratnākara*, *Vādikhaṇḍa* 14.46 (Shastri 1940: 116). According to Hellwig 2009a: 64, the *Vādikhaṇḍa* is among the earliest Indian alchemical works.

54. The wise man performs the work having understood matters in this manner. By means of correctly measuring colors and numbers (*varṇasaṅkhyā*), lead (*nāga*) becomes gold.
55. Now I shall explain the scissor-fixing (*karttarīrasabandhana*) of mercury that arises from the earth, mountains, or creepers, after grasping the subsidiary jewels (*uparatna*).
56. The scissor-fixing of mercury is combined among calamines (*rasaka*) and so on. One should steep these in equal quantities with the urine of a monkey. The wise man bakes this trembling stuff (*capala*) with its flower.[82]
57. After unifying twelve parts of gold and six parts of trembling stuff (*capala*) with sixty-four of mercury (*rasendra*), one should triturate (*mardayet*) them.
58. The wise man puts it into a covered crucible in the form of an udder and then bakes it.[83]
59. When the next day comes, one fires it (√*dhmā*) and makes it bubble (*sphoṭayet*). The mercury looks well bound (*subaddha*) like pure crystal (*sphaṭika*). Then it should be returned to an open crucible (*prakaṭamūṣā*). It is to be repeatedly given into what has been amalgamated (*jīrṇa*). One should not give it into what has not been amalgamated.
60. When mercury devours gold twenty-four times, †like a carnivore, one can put it again into a different mercury.
61. †Marking nut tree (*āruṣkara*), blue lotus blossom (*utpalī*), the sun and the constellations, and what arises from the earth. For these five, well-split, are put together in cold and fire.[84]
62. †It should be placed for three days in the middle of each one.
63. It should be placed (*sthāpitavya*) with a smoke-filled apparatus for three days. †by means of the work of the apparatus it should, with effort, be distilled (*pātitavya*).

63a. One should know that to be pure. It may sweat with sweat from its mouth. And †it is the body of the knowledge of the zodiac [or "one-four-twelve"] in the homes of the sun during March–April.

[82] "Trembling stuff" may refer to bismuth ore or quick-melting tin (Hellwig 2009b: 96) or sometimes quicksilver itself.

[83] On the expression "*mūṣāyāṃ andhayitvā*" see the discussion of "blind crucible (*andhamūṣā*)" by Hellwig 2009b: 48.

[84] The verse is perplexing and irreducibly unmetrical in the first *pāda*.

64. One should mix (*melayet*) all this together with mercury, together with one (*śaśi*), with two-four-three-four (*vedāgniparvalocanaiḥ*). All this should be ground (*peṣayitvā*) †of mercury.
65. The mercury (*rasa*) located on the fire is killed from below and above. The wise man, having ground the mercury (*sūta*) that looks like a hen's egg, †
66. †formerly, according to the rule (*vidhānena*), the piercing one (*vedhin*) should know that the ripe (*pakva*) one is completed (*siddha*), by means of sixteen firings (*puṭa*).
67. If the previous procedure (*saṃskāra*), which is the ultimate, is to be done †twice, the binding should be done (*yojita*) gradually three times.
68. Do the work according to the rules. That mercury essence will be achieved. Aged and then killed, it is aged and killed again.
69. The mercury (*rasa*) that has† gone in and out ten times becomes capable of transmuting ten million times [its own weight] (*koṭivedhin*).

Ratnaghoṣa said:

70. Through effort, having successfully produced the great essence (*mahārasa*) that transmutes ten million times [its own weight] (*koṭivedhin*), it is all for nothing without a body.

[The method of Mārkaṇḍeya]

Nāgārjuna said:

71. I shall be unambiguous and tell you what Mārkaṇḍeya made that produces long life on earth, when there is longevity that is achieved by mercury.[85]

[85] The manuscripts unanimously transmit an anomalous locative, *bhūme*. The authority Mārkaṇḍeya is mentioned in *Suśrutasaṃhitā* (Śārīrasthāna 3.32) as an authority on the formation of a fetus and at the start of the *Carakasaṃhitā* (Sūtrasthāna 1.9) among the ancient authorities of medicine (see also Meulenbeld 1999–2002, vol. IA: 170). The most likely mythic connection is with the story of Cyavana narrated in the Bower manuscript (II.287–312ab), where a recipe called immortality oil (*amṛtataila*), associated with Cyavana, enabled Mārkaṇḍeya to achieve long life. On the mythology of Cyavana, the archetype of artificial longevity, see Leslie 2003.

72. A hundred measures (*pala*) of water of chebulic myrobalan (*abhaya*) fruits, with beleric (*akṣa*) and emblic (*dhātrī*) myrobalans too, are boiled one hundred and eight times. A portion of eight remainders,
73. mixed with ghee, honey, and white sugar (*sitā*), together with ten parts of the three hot substances (*vyoṣa*)[86] and leadwort (*citra*), and the powder of iron (*loha*), prepared with a measure (*pala*) of mercury (*rasa*), is calcinated (*mṛta*).
74. Mica (*abhra*) in the same quantity as †bitumen (*girisuta*),[87] †a leaf of iron (*kāntabhṛṅga*),[88] embelia (*viḍaṅga*), should be well prepared with mercury, with grains (*taṇḍula*) and a lot of Bengal quince (*bilva*).
75 [Make?] †a paste (*kalka*) of white-bark acacia for one month, using an iron vessel, this is the best paste for a daily purified body.
76. †At the time of lying down, Susandhi of the solid wealth, who has the arrogance of a rutting elephant, who is devoted to half of Śiva (Vāmanetra), licks.
77. That prince among men, free from all failings and able to see in all directions with a divine eye, is like the god of love, and the object of love of amorous women.
78. He is like a cloud, full of life, with curly pointed hair; he is like a horse, pure, a true poet and painter.
79. †Motionless like a bull, with the deep sound of a cloud. Like Indra's elephant Suragaja in the world,
80 he has power in the world and lives as long as the moon, the stars and the sun.

[Making pills]

81. And I will tell you another supreme binding pill (*golaka bandha*). By merely consuming this, a man goes beyond old age and death.

[86] The three hot substances are dry ginger, long pepper, and black pepper.

[87] Understood as a synonym for bitumen (*śilājatu*). This substance, *girisuta*, "son of the mountain" is interpreted as talc or ochre or benzoin in some dictionaries, as well as several plants. It could also be the juice of "*giri*" which might be the plant *girikarṇikā*, *Clitoria ternatea* L. or *Symphorema polyandrum* Wight.

[88] This is uncertain; *bhṛṅga* may mean "leaf" in medical literature (Singh and Chunekar 1972: 288). It may also refer to False Daisy (Eclipta prostrata (L.)), though the compound with *kānta* makes this meaning unlikely here.

82. †After making gold, jewels of radiance, together with sulfur, mica, and iron, into the binding pill of equally-aged mercury (*rasendra*) †;
83. †One should smear equal quantities of the five metals onto mercury. And the pill has [the size of] twelve barley-corns.
84. †Even though it is iron, a ground-up pill that is born of mercury has the radiance of a golden moon.[89]
85. †This longevity tonic (*rasāyana*), positioned on a man's chest, and †is a longevity tonic that is immortality, that is a creator.[90]

86–87. †A fragrant paste (*sugandhalepa*) and betel; camphor, saffron, and aloe; sandalwood and musk, cubeb and nutmeg: †these aromatic substances, food and drink, all the places of enjoyment, these are, in turn†,† …

88. †By her[91] who has hair that is curled at the tips, who has dark lotus eyes, of her whose loins are spread out, the heart may be stirred.
89. †She who, during the dark fortnight, may have a showing of the flower because of her youth, she is known as the Crow Woman (*Kākinī*), where longevity is concerned.[92]
90. The fair-eyed woman has a beautiful appearance, when hugging, when something is to be said, when touching, when very beautiful, when lovemaking, and when rubbing.
91. †The termite mound and the disease of the mouth. She who has eyes, ears and so on, and a nose is enjoyed by phlegm, bile, and wind; she is decorated by the virtues of her own nature (*svabhāva*).
92. †When there is dropsy (*jalodara*), there is the cluster fig[93] and the variegated one and the one that is asleep; there is she who seizes (*grāhiṇī*), the abdominal swelling (*gulma*) that has a bad name, swelling of the neck (*gaṇḍamālā*) and stones.

Nāgārjuna said:

[89] Unmetrical.

[90] I am unable to reconstruct the second half of this *vaṃśastha* verse sensibly.

[91] Irregular sandhi and use of the genitive.

[92] There is an equally plausible reading Loving Woman (*Kāminī*) for Crow Woman (*Kākinī*).

[93] The text transmits *udambara*, that has no normal lexical meaning, instead of *udumbara* or *uḍumbavara*, which mean *Ficus racema* L.

93. A man without wrinkles or grey hair, who has been liberated by means of all these, truly lives for three centuries, as powerful as an elephant.[94]

94. †one that has a firm pair (*yuga*) in its mouth is declared of good quality.

95. Now, O best of men, there are balls filled with sixteen items that are fixed firmly in the middle of the mouth. I shall explain which are the qualities that exist in them.

96. That one is not cut by knives, it is not burned by fire. It has the force of wind, great brilliance, and great fame the equal of Śakra.

97–98. In all three worlds, it beguiles the mind, established like the god of love. It is created through desire. In the same way, by merely touching it you create all metals and gold. It has all fractions and no fractions; it can live as long as the moon, sun, and stars.

[Ending the Time of that which is Born][95]

99. I shall once more tell you another binding (*bandha*) that is revered by the best of the gods, the binding called "Ending the Time of that which is Born" (*sūtakālāntaka*) that was declared by the Supreme Being.

99cd–100ab. That one has too few limbs or too many, he is diseased, a hunch-backed dwarf, he has lost his feet, hands and eyes, he's in the grip of decrepitude, and his senses are conquered. He is stupefied, stuttering, mute, and unable to walk.

101-102. He remains like that for the rest of his life, †not free from the three kinds of uncertainty. In this way, by the power of the binding, when he returns he may again become another body (*piṇḍa*). One does not need to think more about this,[96] for the great yoga with five immortalities has been stated in the *Manthānabhairava*.[97]

[94] Verses 91–92 above on medicine appear where the *Rasārṇava* 18.203 omits this material. Ray and Kaviratna 1985: 431n12 noted that verses 18.201–203 of the *Rasārṇava* too were disturbed at this point.

[95] It is possible that this is a passage about abortion or even euthanasia and the promise of a better rebirth. But the text is extremely uncertain.

[96] The phrase *nātra kāryā vicāraṇā* is a regular refrain in the *Rasārṇava*.

[97] The *Manthānabhairava* is a major tāntrika text from the Deccan (Sanderson 2014: 61–62) whose earliest manuscripts are dated to 1180 and 1219 (MS Kathmandu Kesar Library 592 and MS

103. †It was declared again there, by †Vīrākṣasauravat†, in the three worlds, with its living and non-living beings, there is nothing without him.

104. Earth, water, fire, wind, and ether: the mercury (*rasa*) that transmutes millions of times (*koṭivedha*) is to be grasped; the lump (*piṇḍa*) is well united with the oven (*koṣṭha*).

105. The wise man should put the pill (*guṭikā*) in the middle position of each and every one. This is termed the pill (*guṭikā*); †the sixth life is unique.

106. †The six-time copper leaf of the thick lump is very shiny. But above, the measure of a man, there is the inner † circle that is half a man.

107. One should make an oven with four faces. One should make it rest upon it. One should mix together cow's ghee and the great oil in equal quantities.

108. But after filling the cauldron, the virgin should perform the worship of the protectors of the four quarters there, the worship of the groups (*gaṇa*) and the guru.

109. One should offer cakes (*bali*) in the four directions as was declared, as was stated by Śiva.[98] One should blow into the four directions, very slowly.

110–111. †One should know it to be well-heated, and when it is without smoke, having bowed to the moon and the sun, †Having bowed to the divine teacher, one prostrates the body there. One should realize that it is properly burned. One should harness it as the creation.

112. And it becomes an embryo and one should throw down everything once again. One should realize that it has a red color. †One harnesses heats in heat.[99]

113. Wind may lie down at that very spot where there is a lump of meat.

Kathmandu NAK 1/1151/134, respectively; see Dyczkowski 2004: 97 and Goudriaan and Gupta 1981: 55).

[98] There is a particularly clear crux in the MS stemma here between *śivabhāṣita* and *gurubhāṣita*. The worship of a young girl, as instructed by the *Rasārṇava*, is lacking here.

[99] Reading the rare and unlikely noun *teja*.

114ab. †One should cause to give the fact of life to that wandering one that looks like gold.

114cd–115ab. Having done that, a great roar arises that is the sound of Oṃ as worshiped by the gods, then, just as the sun rises at the start of the day, without fail.

115cd–116ab. That Siddha is seen, with divine brilliance, a great body, divine vision, and great strength. He grants all successes to the whole world.[100]

[The Flying-binding]

Ratnaghoṣa said:

117. When it is not possible to make binding that destroys the past, how can one achieve success by means of this body?

Nāgārjuna said:

118. I shall explain to you once more the supreme flying-binding (*khecarabandha*) that leads to equality with the gods merely by being eaten.

[Summary of the remaining verses:]

After Nāgārjuna describes the aging of the color of jewels (121) and other operations, Śiva with saints and his attendants and the Vidhyādharas come to his city (133), and, standing in their flying chariots, decorated with bracelets and tiaras, they view the whole earth (134). He is accompanied by the sound of conches, drums, and the singing and playing of heavenly nymphs decorated with bells. Śiva flies over all obstacles in his path. Nāgārjuna stands before Śiva in reverence (136), and Śiva speaks. He is delighted by Nāgārjuna's devotion and the thousands of difficult and terrible miracles he has performed (137). Śiva congratulates and honors Nāgārjuna as a great sage and declares that he will live in heaven for as long as the moon and stars endure (138). He will live to the end of the kalpa, a representative

[100] Verses 116.1–7 are a stemmatic crux, cleanly dividing MSS A1, J1, Ko1, L, P, and Pa, which omit the section, from A2, G, B, R, Ko2, and U.

of liberation, enjoying pleasures and the maidens of the gods (139). A section of the work on *khecarabandha* ends at this point, and the topic changes back to alchemical operations that will remove the ills of life (141, 158). The fourth section of the *Rasendramaṅgala* ends with assurances of the health benefits of herbal medicines and their value to the physician, but only as long as they are accompanied by mercury and metals that have been operated upon (161–164).

Abbreviations

AcaCu	*Ācārāṅgacūrṇi*, 1941
AcaSi	*Ācārāṅgavṛtti* by Śīlāṅka, 1941
ĀK	*Ānandakandam*, Radhakrishna, 1952
DasCu	*Daśavaikālikacūrṇi*, 1933
RĀ	*Rasārṇava*, Ray and Kaviratna, 1985
RM*ed*	*Rasendramaṅgalam*, Śarmā, 2003
UttCu	*Uttarādhyāyanacūrṇi*, 1933
UttS	*Uttarādhyāyanavṛtti* by Śāntisūri, 1916

Index of Manuscripts

Kathmandu Kesar Library 592: 25
Kathmandu NAK 1/1151/134: 25
Pune BORI 942 of 1891–95: 2

Bibliography

Ācārāṅgacūrṇi. 1941. Ratlam: Rishabhdeo Kesharimal.

Ācārāṅgavṛtti by Śīlāṅka. 1941. Bombay and Mehesana: Agamodaya Samiti.

Ācārya, Yādavaśarma Trivikrama, and Rāmacandraśāstrī Paṇaśīkara, eds. 1939. *Rasâyanakhanda, Fourth Part of Rasaratnākara of Śrī Nitya Nātha Siddha Edited by Vaidya Jâdavjî Tricumjî Âchârya*. Haridāsa Saṃskṛta Granthamālā 95. Banārasa: Caukhambā Saṃskṛta Pustakālaya.

Altekar, A. S., and Vijayakanta Mishra. 1959. *Report on Kumrahar Excavations 1951–1955*. Historical Research Series 3. Patna: K. P. Jayaswal Research Institute.

Āndakandam, See Radhakrishna, 1952

Balbir, Nalini. 1990. "Scènes d'alchimie Dans La Littérature Jaina." *Journal of the European Ayurvedic Society* 2: 149–164.

Balbir, Nalini. 1992. "La Fascination Jaina Pour l'alchimie." *Journal of the European Āyurvedic Society* 2: 134–150.

Beal, Samuel. 1884. *Si-yu-ki, Buddhist Records of the Western World, Translated from the Chinese of Hiuen Tsiang (AD 629).* 2 vols. London: Kegan Paul, Trench, Trübner.

Bhaṭṭācārya, Aśubodha Vidyābhūṣaṇa, and Nityabodha Vidyāratna Bhaṭṭācārya, eds. 1915. *Indrajālavidāsaṅgraha tatra (indrajālaśāstram + kāmaratnam + dattātreyatantram + ṣaṭkarmmadīpikā + siddhanāgārjunakakṣapuṭam.* 3rd ed. Kalikātā: The editors.

Burgess, James. 1876. *Report on the Antiquities of Kāṭhiāwāḍ and Kachh, Being the Result of the Second Season's Operations of the Archaeological Survey of Western India, 1874–75.* Volume 2. Archaeological Survey of Western India. London: W. H. Allen; Trübner; Edward Stanford.

Cantú, Keith. "Alchemical Pilgrimage: Chapter Eight of the Rasaratnākara's Rasāyanakhaṇḍa." Article in the present volume.

Chimpa, Lama, Alaka Chattopadhyaya, and Debiprasad Chattopadhyaya. 1980. *Tāranātha's History of Buddhism in India.* Calcutta: K. P. Bagchi & Co.

Cordier, Palmyr. 1896. *Nāgārjuna Et l'uttaratantra de La Susrutasamhitā.* Études Sur La Médecine Hindou. Antananarivo: J Marivolanitra.

Corless, Roger. 1995. "The Chinese Life of Nāgārjuna." In *Buddhism in Practice*, edited by Donald S. Lopez, 525–531. Princeton, NJ: Princeton University Press.

Cozad, Laurie Ann. 2015. "Nāgas." In *Brill's Encyclopedia of Hinduism*, edited by Knut A. Jacobsen, Helene Basu, Angelika Malinar, and Vasudha Narayanan, 6:72–80. Leiden: Brill. https://doi.org/http://dx.doi.org/10.1163/2212-5019_BEH_COM_000337.

Dās, Sarat Chandra. 1882. "Life and Legend of Nāgārjuna." *Journal of the Asiatic Society of Bengal* 51 (1): 115–120.

Daśavaikālikacūrṇi. 1933. Ratlam: Rishabhdeo Kesharimal.

Desai, Śambhuprasād Haraprasāda. 1968. *Saurāṣṭrano itihāsa* [History of Sauraṣṭra]. Junagaḍh: Soraôh Śikṣan aṇe Sanskrit Sangha.

Dokter-Mersch, Sanne. 2020. "Visual Story-Telling in Text and Image: The Nāga as Inhabitant of the Cosmic Ocean and the Netherworld." In *Framing Intellectual and Lived Spaces in Early South Asia*, edited by Lucas den Boer and Elizabeth A. Cecil, 51–74. Berlin: De Gruyter. https://doi.org/10.1515/9783110557176-003.

Durgāprasād, Pandit, and Kāśīnātha Pāṇduraṅga Parab, eds. 1889. *Mahākaviśrīsoma devabhaṭṭaviracitaḥ kathāsaritsāgaraḥ = The Kathāsaritsāgara of Somadevabhatt.* Mumbai: Nirṇayasāgara Press. https://hdl.handle.net/2027/njp.32101073331967.

Dwivedi, A. V. 2022. "Yakṣa." In *Hinduism and Tribal Religions*, edited by Jeffery D. Long et al., 1793–1794. Dordrecht: Springer. https://doi.org/10.1007/978-94-024-1188-1.

Dyczkowski, Mark S. G. 2004. *A Journey in the World of the Tantras.* Varanasi: Indica Books.

Filliozat, Jean. 1940. "Nâgârjuna Et Agastya, Médecins, Chimistes Et Sorciers." In *Actes Du 20e Congrès International Des Orientalistes*, 228–229. Louvain: Bureaux du Muséon.

Filliozat, Jean. 1951. "Al-Bīrūnī Et l'alchimie Indienne." In *Al-Bīrūnī Commemoration Volume. A.h.362–a.h.1362*, edited by Iran Society, 101–105. Calcutta: Iran Society.

Filliozat, Jean. 1974. *Laghu-Prabandhāḥ. Choix d'articles d'indologie Par Jean Filliozat.* Edited by Collette Caillat, Bruno Dagens, François Gros, and Ludwik Sternbach. Leiden: Brill.

Filliozat, Jean. 1979. *Yogaśataka. Texte Médical Attribué à Nāgārjuna.* Publications de l'institut Français d'indologie 62. Pondichéry: Institut Français d'Indologie.

Filliozat, Jean. 1982. "Al-Bīrūnī and Indian Alchemy." In *Studies in the History of Science in India*, edited by Debiprasad Chattopadhyaya, 1:338–343. New Delhi: Editorial Enterprises.

Forbes, Alexander Kinloch. 1856. *Râs Mâlâ or Hindoo Annals of the Province of Goozerat in Western India.* 2 vols. London: Richardson Co.

Goudriaan, Teun, and Sanjukta Gupta. 1981. *Hindu Tantric and Śākta Literature.* A History of Indian Literature 2.2. Wiesbaden: Harrassowitz.

Granoff, Phyllis E. 1988. "Jain Biographies of Nagarjuna: Notes on the Composing of a Biography in Medieval India." In *Monks and Magicians: Religious Biographies in Asia*, edited by Phyllis Granoff and Koichi Shinohara, 45–67. Oakville, ON: Mosaic Press.

Harrison, Freya, Aled E. L. Roberts, Rebecca Gabrilska, Kendra P. Rumbaugh, Christina Lee, and Stephen P. Diggle. 2015. "A 1,000-Year-Old Antimicrobial Remedy with Antistaphylococcal Activity." Edited by Dianne K. Newman. *mBio* 6 (4). https://doi.org/10.1128/mbio.01129-15.

Hellwig, Oliver. 2009a. "Alchemical Procedures and Their Implications for the Chronology of Medieval *Rasaśāstra*." *eJournal of Indian Medicine* 2: 53–65. https://indianmedicine.nl/article/view/24720/22170.

Hellwig, Oliver. 2009b. *Wörterbuch Der Mittelalterlichen Indischen Alchemie.* Groningen: Barkhuis & University of Groningen, University Library. https://doi.org/10.2307/j.ctt22728hs.

Jinavijaya, Muni, ed. 1933. *śrī merutuṃgācāryaviracita prabaṃdhaciṃtāmaṇi: saṃskṛta graṃtha kā hiṃdī bhāṣāṃtara = Prabandhacintāmaṇi of Merutuṅgācārya Critically Edited in the Original Sanskrit.* 1st ed. Volume 1. Singhi Jaina Series 1. Śāntiniketan: Adhiṣṭhātā, Siṁghī Jaina Jñānapīṭha.

Jinavijaya, Munirāja. 1920. *Śrīsomaprabhācāryaviracitaḥ kumārapālapratibodhaḥ. Kumārapāla-Pratibodha of Somaprabhāchārya, Edited with English Introduction, Samskrit Prastāvanā and Pariśiṣṭa.* Baroda: Central Library.

Kano, Kazuo. 2020. "References to the Preservation of Sanskrit Manuscripts in Medieval Tibetan Monastries: gCung Ri Bo Che and sPos Khang." *Journal of Indian and Buddhist Studies (Indogaku Bukkyogaku Kenkyu)* 68 (2): 913–907. https://doi.org/10.4259/ibk.68.2_913.

Kapadia, Hiralal Rasikdas. 1941. *A History of the Canonical Literature of the Jainas.* Shree Shwetambar Murtipujak Jaona Boarding (Ahmedabad) Series 17. Gopipura, Surat: The Author.

Karambelkar, V. W. 1952. "The Problem of Nagarjuna." *Journal of Indian History* 30 (88): 21–34.

Khandelavala, S. N. 1996. *Siddhanāgārjunatantram (siddhanāgārjunakakṣapuṭam) hindīvyākhyopetam.* Vārāṇasī: Caukhambā Surabhāratī Prakāśana.

Leslie, Julia. 2003. *Authority and Meaning in Indian Religions. Hinduism and the Case of Valmiki.* Aldershot: Ashgate. https://doi.org/10.4324/9781315198439.

Lindtner, Christian. 1982. *Nagarjuniana: Studies in the Writings and Philosophy of Nāgārjuna.* Copenhagen: Akademisk Forlag.

Lopez, Donald S. Jr. 2019. *Seeing the Sacred in Samsara: An Illustrated Guide to the Eighty-Four Mahasiddhas.* Boulder, CO: Shambala.

Mabbett, Ian. 1998. "The Problem of the Historical Nagarjuna Revisited." *Journal of the American Oriental Society* 118: 332–346. http://www.jstor.org/stable/606062.

Maithrimurthi, Mudagamuwe. 1999. *Wohlwollen, Mitleid, Freude Und Gleichmut: Eine Ideengeschichtliche Untersuchung Der Vier Apramāṇas in Der Buddhistischen Ethik Und Spiritualität von Den Anfängen Bis Hin Zum Frühen Yogācāra.* Stuttgart: Franz Steiner Verlag.

Matsunaga, Yukei. 1985. "On the Date of the Mañjuśrīmūlakalpa." In *Tantric and Taoist Studies in Honour of R. A. Stein*, edited by Michel Strickmann, 3:882–894. Brussels: Institut Belge des Hautes Études Chinoises.

Mehta, Mohanlal, and K. Rishabh Chandra. 1970–1972. *Prakrit Proper Names.* Edited by Dalsukh Malvania. 2 vols. Ahmedabad: L. D. Institute of Indology.

Meulenbeld, Gerrit Jan. 1999–2002. *A History of Indian Medical Literature.* . 5 vols. Groningen: E. Forsten.

Mitra, Rajendralal. 1880. *A Catalogue of Sanskrit Manuscripts in the Library of the Maharaja of Bikaner.* Calcutta: Government of India.

Mookerji, Bhudeb. 1926–1938. *Rasa-Jala-Nidhi or Ocean of Indian Chemistry and Alchemy. Compiled in Sanskrit with English Translation by the Author.* 5 vols. Calcutta: The author.

Newman, William R. 2018. *Newton the Alchemist: Science, Enigma, and the Quest for Nature's "Secret Fire."* Princeton, NJ, and Oxford: Princeton University Press.

Obermiller, E. 1931–1932. *The History of Buddhism in India and Tibet. The Jewelry of Scripture, by Bu Ston, Translated from Tibetan.* 1st ed. 2 vols. Leipzig: Harrassowitz.

Ollett, Andrew. 2018. "Sātavāhana and Nāgārjuna: Religion and the Sātavāhana State." *Journal of the International Association of Buddhist Studies* 41: 421–472. https://doi.org/10.2143/JIABS.41.0.3285748.

Pādhye, Haṇamantaśāstrin Kṛṣṇaśāstrin, ed. 1894. *Srīmad-vṛnda-praṇīto vṛndamādhavāparanāmā siddhayogaḥ śrīkaṇṭhadattaviracitavyākhyākusumāvalyākhyaṭīkāsametaḥ.* Ānandāśramasaṃskṛtagranthāvaliḥ 27. Puṇe: Mahādeva Cimaṇājī Āpaṭe for the Ānandāśramamudraṇālaya.

Plessner, M. 1954. "The Place of the Turba Philosophorum in the Development of Alchemy." *Isis* 45 (4): 331–338. https://doi.org/10.1086/348355.

Radhakrishna, S. V., ed. 1952. *Anandakandam, Edited with Translation in Tamil, and Introduction in Tamil and Sanskrit.* Volume 15. TMSSM Series 15. Thanjavur: Tanjore Maharaja Serfoji's Saraswati Mahal Library.

Rasaratnasamuccaya, See Vidyābhūṣaṇa and Vidyāratna, 1927

Rasārṇava, See Ray and Kaviratna, 1985

Rasendramaṅgalam, See Śarmā, 2003

Rāy, Praphulla Chandra. 1902–1909. *A History of Hindu Chemistry from the Earliest Times to the Middle of the Sixteenth Century, A.D. With Sanskrit Texts, Variants, Translation and Illustrations.* Volumes 1–2. Calcutta and London: Prithwis Chandra Rāy; London & Norgate.

Ray, Praphulla Candra, and Hariś Candra Kaviratna, eds. 1985. *Rasārṇava* . Reprint of the 1910 edition. Calcutta: The Asiatic Society.

Robinson, James B. 1979. *Buddha's Lions: The Lives of the Eighty-Four Siddhas: Caturaśīti-Siddha-Pravṛtti by Abhayadatta. Translated into Tibetan as Grub Thob Brgyad Cu Rtsa Bzhi'i Lo Rgyus by sMon-Grub Shes-Rab. Translated into English.* Tibetan Translation Series 10. Cazadero, CA: Dharma Publishing.

Rongxi, Li, and Albert A. Dalia. 2002. *Lives of Great Monks and Nuns: The Life of Aśvaghoṣa Bodhisattva Translated from the Chinese of Kumārajīva by Li Rongxi; the Life of Nāgārjuna Bodhisattva Translated from the Chinese of Kumārajīva by Li Rongxi; Biography of Dharma Master Vasubandhu Translated from the Chinese of Paramārtha by Albert A. Dalia; Biographies of Buddhist Nuns Translated from the Chinese of Baochang by Li Rongxi; the Journey of the Eminent Monk Faxian / Translated from the Chinese of Faxian by Li Rongxi [Taishō vol. 50, Nos. 2046, 2047, 2049, 2063 and v. 51, No. 2085].* Volume 76: III–VII. BDK English Tripitaka. Berkeley: Numata Center for Buddhist Translation; Research. https://bdkamerica.org/download/1878.

Roşu, Arion. 1992. "Alchemy and Sacred Geography in the Medieval Deccan." *Journal of the European Āyurvedic Society* 2: 151–157. https://doi.org/10.5281/zenodo.10070412.

Ruegg, David Seyfort. 1981. *The Literature of the Madhyamaka School of Philosophy in India.* Wiesbaden: Harassowitz.

Sachau, Edward C. 1910. *Alberuni's India: An Account of the Religion, Philosophy, Literature, Geography, Chronology, Astronomy, Customs, Laws and Astrology of India about AD 1030.* 2 vols. London: Kegan Paul, Trench, Truebner; Co.

Sanderson, Alexis. 2009. "The Śaiva Age: The Rise and Dominance of Śaivism During the Early Medieval Period." In *Genesis and Development of Tantrism*, edited by Shingo Einoo, 41–349. Tokyo: University of Tokyo.

Sanderson, Alexis. 2014. "The Śaiva Literature." *Journal of Indological Studies* 24–25 (2012–2013): 1–113.

Sankalia, Hasmukh Dhirajlal. 1938. "The Earliest Jain Sculptures in Kāthiāwār." *Journal of the Royal Asiatic Society of Great Britain and Ireland* 3: 426–430. https://www.jstor.org/stable/25201741.

Śarmā, Hari Śaṅkara. 2003. *Rasendramaṅgalam of Nāgārjuna Edited with Aihore Hindī Vimarśa, Bhāvānuvāva and English Translation and Notes (Chapters 1–4)*. Jaikrishnadas Āyurveda Series 119. Vārāṇasī: Chaukhambha Orientalia.

Shah, A. M., R. G. Shroff, and M. N. Srinivas. 1958. "The Vahīvancā Bāroṭs of Gujarat: A Caste of Genealogists and Mythographers." *The Journal of American Folklore* 71 (281): 246–276. https://doi.org/10.2307/538561.

Sharma, Har Dutt. 1939. *Descriptive Catalogue of the Government Collections of Manuscripts Deposited at the Bhandarkar Oriental Research Institute*, Volume 16, Part 1: *Vaidyaka*. Descriptive Catalogue of Manuscripts in the Government Manuscripts Library. Pune: Bhandarkar Oriental Research Institute.

Sharma, Hari Śaṅkar. 1999. "Rasendramangalam, Text and Translation." *Indian Journal of History of Science* 34 (3–4): S1–27, S29–66.

Sharma, Priya Vrat. 1993. "A Fragment of the Lauhaśāstra of Nāgārjuna." *Indian Journal of History of Science* 28 (1): 35–50.

Sharma, Priya Vrat, ed. 1994. *Cakradatta (Text with English Translation: A Treatise on Principles and Practices of Ayurvedic Medicine)*. Kashi Ayurveda Series 17. Varanasi, Delhi: Caukhambha Orientalia.

Shastri, Jivaram Kalidas, ed. 1940. *Ruddhikhanda Vadikhanda from: Rasa Ratnakar Composed Shri Parvati Putra Nityanath Siddha*. Rasashala Granthamala 9. Gondal: Rasashala Aushadhashram.

Simioli, Carmen. 2013. "Alchemical Gold and the Pursuit of the Mercurial Elixir." *Asian Medicine* 8 (1): 41–74. https://doi.org/10.1163/15734218-12341289.

Singh, Thakur Balwant, and K. C. Chunekar. 1972. *Glossary of Vegetable Drugs in Bṛhattrayī*. Varanasi: Chowkhamba Sanskrit Series Office.

Tawney, C. H. 1901. *The Prabandhacintāmaṇi or Wishing-Stone of Narratives. Composed by Merutuṅga Ācārya. Translated from the Original Sanskrit*. Calcutta: The Asiatic Society.

Tawney, C. H., and N. M. Penzer. 1924–1928. *The Ocean of Story, Being C. H. Tawney's Translation of Somadeva's Kathā Sarit Sāgara (or Ocean of Streams of Story) Now Edited with Introduction, Fresh Explanatory Notes and Terminal Essay ... in Ten Volumes*. 10 vols. London: Chas. J. Sawyer. https://tinyurl.com/2p98dd5a.

Thompson, Gordon. 1992. "The Bāroṭs of Gujarati-Speaking Western India: Musicianship and Caste Identity." *Asian Music* 24 (1): 1–17. https://doi.org/10.2307/834448.

Tucci, Giuseppe. 1929. *Pre-Diṅnāga Buddhist Texts on Logic from Chinese Sources*. Volume 49. Gaekwad's Oriental Series. Baroda: Oriental Institute.

Uttarādhyāyanacūrṇi. 1933. Ratlam: Rishabhdeo Kesharimal.

Uttarādhyāyanavṛtti by Śāntisūri. 1916. Bombay: Devchand Lalbhai Series.

Vidyābhūṣaṇa, Āśubodha, and Nityabodha Vidyāratna, eds. 1927. *[Rasaratnasamuccaya with a Commentary]*. Kalikātā: The editors.

Virji, Krishnakumari J. 1952. *Ancient History of Saurashtra (Being a Study of the Maitrakas of Valabhi V to VIII Centuries A.D.)*. Indian History and Culture Series / Konkan Institute of Arts and Sciences 1. Bombay: Konkan Institute of Arts; Sciences.

Walleser, Max Gebhardt Lebrecht. 1923. "The Life of Nagarjuna from Tibetan and Chinese Sources." Translated by Arthur Probsthain. *Asia Major* Hirth Anniversary Volume: 421–455.

Walleser, Max Gebhardt Lebrecht. 1979. *The Life of Nāgārjuna from Tibetan and Chinese Sources*. Delhi: Nag Publishers.

Walser, Joseph. 2002. "Nāgārjuna and the Ratnāvalī. New Ways to Date an Old Philosopher." *Journal of the International Association of Buddhist Studies* 25 (1–2): 209–262. https://www.academia.edu/39303593/.

Walser, Joseph. 2005. *Nagarjuna in Context: Mahayana Buddhism and Early Indian Culture*. New York: Columbia University Press. https://doi.org/10.7312/wals13164.

Walser, Joseph. 2014. "Nagarjuna: The Great Philosopher." In *Buddhists: Understanding Buddhism Through the Lives of Believers*, edited by Todd Lewis, 39–51. Lived Religions. Chichester: Wiley.

Wenzel, Heinrich. 1886. *Suhṛllekha: Brief Des Nāgārjuna an König Udayana Aus Dem Tibetischen Übersetzt*. Leipzig: Voss' Sortiment.

Westerhoff, Jan Christoph. 2022. "Nāgārjuna." In *The Stanford Encyclopedia of Philosophy*, edited by Edward N. Zalta, Summer. Metaphysics Research Lab, Stanford University. https://plato.stanford.edu/archives/sum2022/entries/nagarjuna/.

White, David Gordon. 1996. *The Alchemical Body: Siddha Traditions in Medieval India*. Chicago: University of Chicago Press.

Wilberforce-Bell, H. 1916. *The History of Kathiawad from the Earliest Times*. London: William Heinemann.

Wiles, Royce. 2006. "The Dating of the Jain Councils: Do Scholarly Presentations Reflect the Traditional Sources?" In *Studies in Jaina History and Culture: Disputes and Dialogues*, edited by Peter Flügel, 61–85. London: Routledge. https://doi.org/10.4324/9780203008539-12.

Wilson, Clare. 2015. "Anglo-Saxon Remedy Kills Hospital Superbug MRSA." *New Scientist*, March. https://tinyurl.com/lhxgg9w.

Wujastyk, Dominik. 1984. "An Alchemical Ghost: The *Rasaratnākara* by Nāgārjuna." *Ambix* 31: 70–83. https://doi.org/10.1179/amb.1984.31.2.70.

Wujastyk, Dominik. 2023. "The Surviving Manuscripts of the *Rasendramaṅgala* of Nāgārjuna Siddha." https://doi.org/10.5281/zenodo.6486033.

Yamano, Chieko. 2008. "Nagarjuna and Satavahana." *Journal of Indian and Buddhist Studies* 56 (2): 1157–1163. https://www.academia.edu/56579231.

Yamano, Chieko. 2009. "Śrīparvata as Nāgārjuna's Abode: At the Confluence of Legend, History, and Geography." *IBK* 57 (3): 1246–1252. https://www.academia.edu/51964599.

Ye, Shaoyong. 2019. "Nāgārjuna." In *Brill's Encyclopedia of Buddhism*, edited by Jonathan Silk, Richard Bowring, Vincent Eltschinger, and Michael Radich, 2:335–347. Leiden: Brill. https://www.academia.edu/39717987.

Ye-shes-rdo-rje, Klong-chen. 2005. *Nagarjuna's Letter to a Friend: With Commentary by Kangyur Rinpoche*. Ithaca, NY: Snow Lion Publications.

Yün-Hua, Jan. 1970. "Nāgārjuna, One or More? A New Interpretation of Buddhist Hagiography." *History of Religions* 10 (2): 139–155. https://doi.org/10.1086/462624.

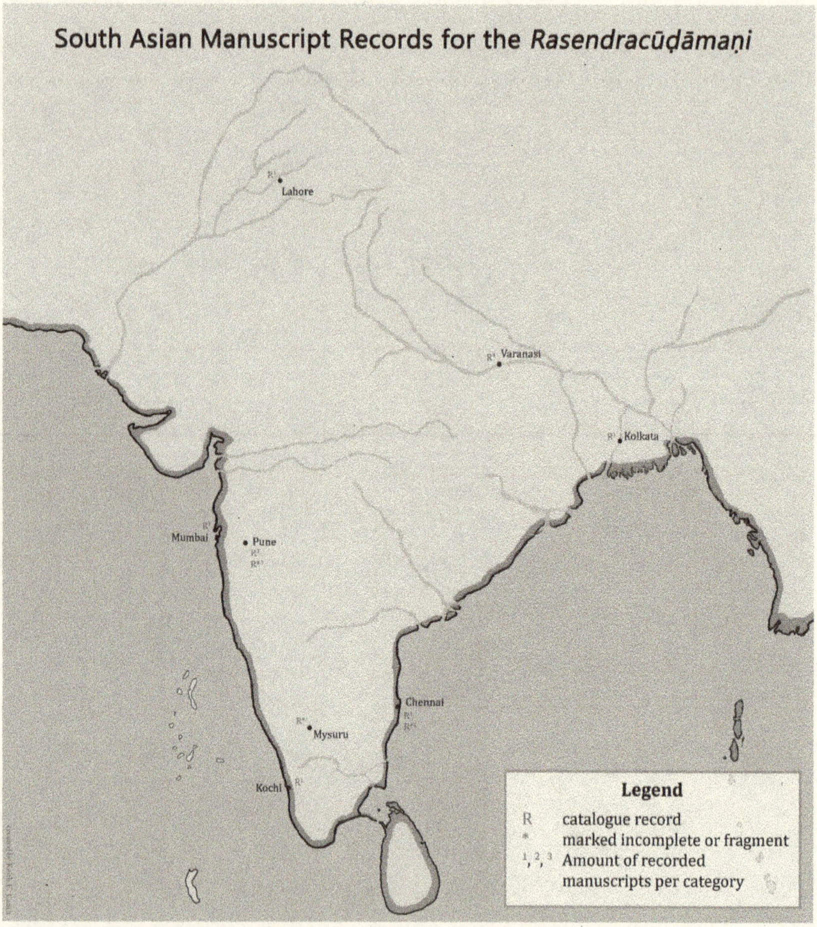

Map 4.1 South Asian manuscript records for the *Rasendracūḍāmaṇi*.

4

The Alchemical Laboratory

Chapter 3 of Somadeva's *Rasendracūḍāmaṇi*

Patricia Sauthoff

In alchemical literature, we find three distinct but related alchemical spaces. The mercurial temple (*rasamaṇḍapa*), found in the *Rasaratnasamuccaya*, *Rasaratnākara*, *Ānandakanda*, and *Rasārṇava*, differs from the laboratory (*rasaśālā*), described in the *Rasendracūḍāmaṇi*, *Rasaratnasamuccaya*, and *Ānandakanda*, and the hut (*kuṭi*),[1] a unique feature of the *Ānandakanda*.

Descriptions of the alchemical laboratory (*rasaśālā*) have remained consistent throughout the centuries. This largely stems from the borrowing of verses that describe the *rasaśālā* from its original source, the twelfth- to thirteenth-century *Rasendracūḍāmaṇi*, or *Crest-Jewel of the Lord of Essences* by Somadeva. This is the first and most comprehensive description of the *rasaśālā* found in alchemical literature.[2] The *Rasa Jala Nidhi*, a twentieth-century compilation, rearranges several of the verses, and this reordering has been carried on in subsequent editions of the *Rasendracūḍāmaṇi*. The sixteenth-century *Rasaratnasamuccaya*[3] also borrows from Somadeva's *Rasendracūḍāmaṇi*[4] with little deviation. Various editions of the *Rasaratnasamuccaya* present the *Rasendracūḍāmaṇi*'s *rasaśālā* chapter with slight differences in the sequence of verses, occasional changes in individual words, or small grammatical adjustments. As noted by Bendixen, the *Rasaratnasamuccaya* changes both the content and verse order that it has borrowed from the *Rasendracūḍāmaṇi* throughout the text. In fact, these changes are so prevalent throughout the *Rasaratnasamuccaya*, Bendixen

[1] *Rasaratnākara* 2.4.81 mentions the *kuṭi* in the mantra *oṃ kuṭijātānām amṛtaṃ kuru kuru svāha*. This mantra also appears at *Ānandakanda* 1.15.102.

[2] However, the eleventh-century *Rasārṇava* (2.107–108) describes a parallel structure with similar features that it calls *maṇḍapa*, and does not seem to differentiate from the ritual space described in the verses before. See the translation of the second chapter of the *Rasārṇava* by White in this volume.

[3] See Meulenbeld 2000a: 670–671 for a discussion of the possible dating of the *Rasaratnasamuccaya*.

[4] White 1998: 488nn221–222.

Patricia Sauthoff, *The Alchemical Laboratory* In: *Indian Alchemy*. Edited by: Dagmar Wujastyk, Oxford University Press. © Oxford University Press 2025. DOI: 10.1093/9780197805831.003.0005

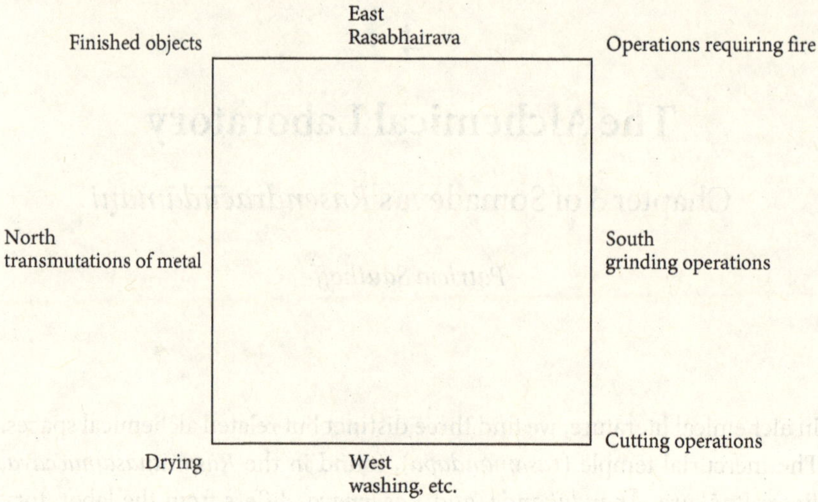

Figure 4.1 Layout of the laboratory.

calls it a sort of new edition of the *Rasendracūḍāmaṇi*.[5] Aside from these very minor changes, the chapter remains in full.

This means the layout of the laboratory, with the deity Rasabhairava[6] placed in the east and so on (see Figure 4.1), deviates little across texts.[7] However, the *Rasendracūḍāmaṇi* also contains enough ambiguity in its descriptions of the implements within the lab that we cannot be positive the bottles, crucibles, and sieves described therein remain identical to those in use today. Several items, such as the *kaṭatrāṇī*,[8] *kṣudrakṣipra*,[9] *pālikā*, and

5 Bendixen 1990: 164.

6 *Rasārṇāva* 2.110–116 describes Rasabhairava with the same characteristics and mantra as the tantric deity Svacchandabhairava as he appears in both the seventh-century *Svacchanda Tantra* and the eighth-century *Netra Tantra*. White 1988: 178–179, 442nn57–59, 443n62. *Ānandakanda* 1.21 elaborates on the Śaiva tantric deities Bhairava (1.21.12–16), Mṛtyuñjaya (1.21.17.1), and Vaṭuka (1.21.17.1–21).

7 See the layout described in *Rasārṇava* chapter 2, verses 107–108 (see White, in this volume), which is very similar, but slightly changes the positioning of activities.

8 Meulenbeld 2000b: 707n274 translates the term as a mortar "according to the Hindī commentary," while Nath 2017: 196 describes *kaṭatrāni* as scissors made of stone and metal for cutting sheets of metal and leather. Dole 2015: 263 agrees with Nath in his description of *kaṭatrāni* as "instruments to cut sheets of metals (or pieces of soft leather)." Hellwig 2009: 196 suggests *kaṭatrāni* could also be a *dvandva* compound of *kaṭatra* and *āṇi*, meaning a mortar and some sort of rod or skewer. He notes that the purpose of *kaṭatra* remain unclear.

9 A sharp-edged knife according to Meulenbeld 2000b: 707n275; small pearl shells according to Nath 2017: 197; Mookerji 1927: 15, reads *kṣudrakṣipra* as two items, *kṣudra* being a cooking spud, or *khunti* in Bengali and *kṣipra* as forceps. Dole 2015: 264 describes *kṣudrakṣipra* as small-sized conches.

karṇikā,[10] have been described in several very different ways with little consensus among commentators and translators. These variations aside, we find the early laboratory remained intact over the centuries.

The *rasaśālā* should be located in a pleasant garden, filled with medicinal plants and fresh water, and away from dangers. Unfortunately, the *Rasendracūḍāmaṇi* does not go into detail about the plants in the garden. The *Ānandakanda* devotes sixteen verses in its second chapter to the *rasaśālā* garden.[11] The medicines for revitalization therapies made within the lab are applied within the hut under the watch of the alchemist. This implies that the hut, laboratory, and temple are located on the same property.[12] The alchemist is to build his *rasaśālā* in a place with good government and free from harm, on even ground, and enclosed with a wall that has an iron gate. The gate and wooden locks described in the *Ānandakanda* indicate a permanency for the buildings. The *Ānandakanda* details the laboratory's garden full of orange, pomegranate, citrus fruit, palmyra, date, bakula (here vakula), coconut, tamarind, trumpet flower, Ashoka, rose apple, jaka, shala, kumkum, wood apple, betel nut, pine, gooseberry, ficus, sacred garlic pear, mahua, wild mango, and screw pine trees as well as six kinds of jasmine, pepper, arni, and seven types of lotus flowers. The garden attracts geese, Indian cranes, ducks, and cakra birds, with lakes and cold water all around.[13] With such lush surroundings and three permanent or semi-permanent structures devoted to practice and therapy rather than daily life, like the *Rasendracūḍāmaṇi*, the *Ānandakanda* and other texts confirm that the alchemist enjoys a certain affluence.[14]

[10] Both are said to be "probably cutting instruments" according to Meulenbeld 2000b: 707n276 and Dole 2015: 264. Nath 2017: 197 describes *pālikā* as a sharp-edged blade and *karṇika* as "ear-ring shaped instruments in small size [sic] for cutting and scraping"; Mookerji 1927: 16 calls *pālikā* a small dagger and a *karṇika* a *kurni* or scraper in Bengali. Citing *Rasaratnasamuccaya* 9.55, Hellwig 2009: 166 equates *pālikā* with *ālavāla*, a basin for water that sits around the root of a tree. *Rasaratnasamuccaya* 9.50 describes the *pālikāyantra* as an iron cup (*caṣaka*) with a long vertical handle, bent at the end and used for the digestion (*jāraṇa*) of sulfur.

[11] Meulenbeld 2000a: 592 gives the twelfth- to thirteenth-centuries as the upper limit of the *Ānandakanda*, but I believe it may be at least several centuries later.

[12] *Rasaratnasamuccaya* 6.13–14 states that the *rasamaṇḍapa* be built near the *rasaśālā*; White 1996: 488n221.

[13] *Ānandakanda* 1.2.24–30– *nāraṅgadāaimaḍahajambīraphalapūrake /*
tālahintālavakulanārikelāmlapālale // (26)
aśokajambūpanasasālapuṃnāgamaṇḍite /
kapitthapūgasaraladevadārusubilvake // (27)
dhātrīnyagrodhavaraṇamadhūkāmrātamaṇḍite /
ketakīmallikājātīyūthikāmālatīyute // (28)
māgadhīkundakuravatilake śatapatrake /
kumudotpalakalhārapuṇḍarīkārbujotpalaiḥ // (29)
haṃsasārasakāraṇḍacakravākavirājitaiḥ /
sarobhiḥ śītalajale paritaḥ pariveṣṭite // (30)

[14] Additionally, the cost of materials and the secrecy necessary for alchemical operations indicate that the alchemist was probably a landowner outside of or on the outskirts of an urban center.

Following on from its lists of the laboratory's inventory, chapter 3 of the *Rasendracūḍāmaṇi* also lists the persons involved in alchemical operations, describing the qualities of the alchemists (*rasavaidya*), experts (*vaidya*), and assistants (*paricāraka*) who work together. Knowledge of alchemical works and mantras are vital for alchemists and experts while assistants are defined by their moral characteristics, such as purity, bravery, honesty, and piety to Śiva. Similarly, those who utilize mercury must be pure in thought and action for therapies to prove successful. These descriptions parallel those first set out in *Rasārṇava* chapter 2.

Notes on the Text and Translation

For this translation I have relied on the unpublished edition completed by Arion Roşu prior to his death in 2007.[15] Roşu based his edition on five manuscripts and two published editions. Chapter 3 contains variants from four texts: Pune, 153/A1882–1883; Paris, Bibliothèque Nationale, Cordier 1934, 1254; Calcutta Sanskrit College, X. No. 54;[16] and the published edition of Yadavji Trikamji Ācārya, Rasendra Chudamani, Motilal Banarsidass, Lahore, 1932. Further manuscripts used by Roşu in his critical edition include: University of Pennsylvania, Ms. Coll. 390, Item 789; Paris, Bibliothèque Nationale, Cordier 1934, 1221; Varanasi, Sarasvati Bhavan XII, Nr. 45067; and auxiliary notes in Trikamji. Several additional manuscripts appear to have not been consulted, namely: Pune, BORI, 939/1884–1887.

As I noted above, recent published editions of the *Rasendracūḍāmaṇi* and *Rasaratnasamuccaya* present the text in a different order than that used here. Roşu's edition solves several difficulties and ambiguities within the texts as published though it remains only a draft and not a complete critical edition. Roşu demonstrates that abrupt changes in subject matter are due to the misordering that has occurred after the publication of the *Rasa Jala Nidhi*. This renders the titles added by Miśra in his 2009 edition unnecessary. I have made emendations where necessary.

[15] Barois 2016.

[16] See Meulenbeld 2000b: 704n203 for this reference. Roşu notes only C in his annotation to his edition.

The *Rasendracūḍāmaṇi* is an early South Asian alchemical text that is referenced or borrowed from widely in later alchemical works.[17] Its sixteen chapters begin with devotional praises of Śiva, Rāma, Rasamahābhairava, and Rasāṅkuśī (the goddess of mercury) and describes detailed worship and initiatory rites for alchemical practitioners. Further, the text pays homage to the Kaulika tradition. This demonstrates how the text borrows from the argot of earlier Bhairava Tantras[18] while it focuses its practices on worldly rather than salvific aims. In addition to its praise of the deities, the *Rasendracūḍāmaṇi* features definitions of technical terms, descriptions of alchemical apparatuses (*yantras*), descriptions of divine and medicinal plants and minerals, the extraction of essences, characterization of gems and metals, and the curative properties of mercury and other substances.

As Roşu notes, a historical reconstruction of the development of alchemical practices is only in its infancy. However, with its rich and thorough descriptions, the *Rasendracūḍāmaṇi* offers cultural features that can illuminate the incomplete history of Indian alchemy.[19]

[17] Texts that quote the *Rasendracūḍāmaṇi* include the *Rasamañjarī, Yogaratnākara, Capalnirṇaya, Rasakāmadhenu, Rasamitra, Rasaratnākara, Rasaratnasamuccaya, Rasasindhu*, and others. See Meulenbeld 2009a: 709 for a more complete list.

[18] Roşu 1997: 410.

[19] Roşu,1997: 408–409.

Rasendracūḍāmaṇi, Chapter 3

1–2. One should set up the splendid laboratory (*rasaśālā*), equipped with the various instruments in a place free of affliction, containing all medicinal plants, beautiful, having a well, and in the very auspicious direction of the Lord of Yakṣas (Kubera/north) or the Thousand Eyed (Indra/east), surrounded by an enclosure.

2–5. He should place Rasabhairava in the lab's eastern portion, operations requiring fire in the direction of Agni (southeast) [grinding and so on], operations using stone in the direction of Yama (south), cutting instruments in the direction of Nairṛti (southwest), washing etc. in the direction of Vāruṇa (west), in the direction of Vāyu (northwest) the drying, and thus the transmutation of metals (*vedhakarma*) in the north. He should store finished products in Īśa's corner (northeast). The collection of all the objects is to be done for the processing of mercury.

5–7. He should also assemble all the various instruments: a splendid apparatus for the extraction of mercury; an oven; an underground stove; a portable oven; many water jugs; a pair of bellows and likewise a pair of tubes made of metal and bamboo; gold, iron, brass, copper, and stone pots; and pots made of leather.

8–14. He should also assemble all various instruments, a churning stick, pestle, mortar, and tapered (*droṇi*) or cylindrical (*vartula*) containers; an iron mortar and the corresponding pestle [that can be] heated; a sieve with 1,000 small holes for the sieving of material; a cloth sieve; and a threshing sieve made of stone or metal. The shapes of the three named sieves are now described. A sieve made of bamboo strips tied with twine, this is always known to be suitable for straining thick liquids. Another sieve for the purpose of sieving powders is made of young bamboo and the bark of the karnikara tree, seemal tree, and the coral tree. A final sieve, covered in a goat hide, four finger lengths wide and one cubit in circumference with a base of stretched cloth and horse hair. With it, one may sieve to retain the finest dust.

15. [There should be] clay, chaff, cotton, dried and shaped cow dung, and dung powder from forest cows for a crucible, as well as three kinds of medicine, made of mineral, animal, and plant.

16–24ab. Charcoal, powdered cow dung, rock sugar, and white sugar are used. Charcoals are called *pāvakocchiṣṭa*, *aṅgāra*, and *kokilā*. *Aṅgāra*s are charcoals separated from fire, and *kokilā*s are coals that are extinguished without water.[20] The names for dried cow dung are thus: *piṣṭaka*, *chagaṇa*, *chāṇa*, *upala*, *utpala*, and *karaṇḍopalasārī*. Bottles and cups made of glass, iron, clay, and shell, the bottles named *champikā*, *siddhā*, *golā*, and *karaṇḍikā* are used and the synonyms for cups are *caṣaka*, *kaṭorī*, and *vāṭikā*, *dhoṭika*, *kacolī*, and *grāhikā*.[21] Sifting baskets etc., bamboo dishes, small shells, and conches, as well as sharp-edged knives, pots, and anything else, a ladle, a wick-shaped knife, and knives for cutting vegetables, that which is for the cleaning of the lab and for the processing of mercury [are used]. One should assemble there anything else that is used, having consecrated [the instruments] with the highest mantra dedicated to Rasāṅkuśī. Otherwise, Bhairava seizes their inherent power.

24bc–25. Experts in the two alchemical compilations and skilled practitioners who know the lexicons and are conversant with the languages of all countries should be brought together by the aspiring alchemists. One should always recite [the mantra of] Aghora at the conclusion of the roasting of mercury.

26–28. One who is loyal to the alchemist is equipped, pure, brave, and most powerful, virtuous, honest, knowledgeable, and honors the long-haired Śiva, compassionate and lotus-handed, is employed by the alchemist. He would be considered one who has the nectar hand: the hand marked by the flag, pitcher, lotus, fish, and bow, [with] well-developed marks and lines at the base of the hand.

[20] See *Rasakāmadhenu* 1.2.22.3.

[21] Bottles and cups with the same names are used in modern rasashastra practice. Here, a *champikā* is a long flat bottle, a *siddha* may be a narrow-necked bottle, a *golā* is a small round bottle, and a *karaṇḍikā* is a big storage bottle. A *caṣaka* is a large cup of the size of a trophy, a *kaṭorī* is a medium-sized bowl (like a bowl for cereal), a *vāṭikā* is a small bowl, a *dhoṭika* is a big bowl (like a salad bowl), a *kacolī* is a shallow plate-like bowl, and a *grāhikā* is a small bowl. I would like to thank Dr Priyanka Chorge for these identifications. Whether modern denominations correspond to their historic counterparts is uncertain.

29. The foreign, uncompassionate, greedy, and unsuitable physician who has a burnt hand or dark lines should be avoided.
30. Those who know the mantras to ward off demons are entrusted with preparing the stores. They are most powerful, resolute, red-eyed, dark-bodied, are experts on terrifying demons and employed in the making of oblations.

31–32. Those who are free from greed, truth-speakers, who honor gods and brāhmaṇas, who are restrained, and who eat suitable foods, should be engaged in elixir regimen (*rasāyana*).[22] They are wealthy, generous, equipped with all apparatuses, always delight in the words of the guru, and are well-versed in metallurgy.

33. They know the names of various medicines, are pure, free of deceptions, know various languages and subjects,[23] and are versed in the procurement of medicinal herbs.
34. Mercury succeeds for those who are pure, speak the truth, are faithful, thoughtful, and whose minds have abandoned doubt, not otherwise.
35. After the mercury has been perfected with the eighteen procedures, the best of adepts should state "Oh, the mercury is spoiled,"[24] and should partake of the mercury elsewhere.
36. The mortal will become one who is perfected through mercury (*rasasiddha*), who gives and enjoys but is not a beggar. Freed from ageing, honored by kings, he has heavenly splendor and is always happy.

Bibliography

Āchārya, Vaidya Jādavjī Tricumjī. 1939. *Rasaratnākara*, of Nitya Nātha Siddha. Benares: Caukhamba Sanskrit Office.

Ānandakanda of Mahābhairava. See Radhakrishna Shastri 1952.

Bapta, Krsnaravasarman Vinayaka. 1890. *Rasaratnasamuccaya of Vāgbhaṭa*. Poona: Anandasrama Press.

[22] See Maas 2017: 66–84 for the difficulties in translating the term *rasāyana*, and Dagmar Wujastyk in Chapter 1 in this volume.

[23] Here *viṣaya* can also mean regions in addition to subjects. Therefore, the sentence could also read "knows the languages of different regions."

[24] *daśāṣṭakriyayā siddhe rase 'sau sādhakotamaḥ / hā raso naṣṭa ity uktvā sevetānyatra taṃ rasam* // The term translated as "spoiled" above is *naṣṭa*. This could also be understood as denoting "dead", in allusion to the "killing" (*māraṇa*) of mercury as the final result of the procedures. This verse is different in the print edition, which reads "*mahāraso 'yam*", "This is the great mercury."

Barois, Christèle. 2016. "The Arion Roşu Collection of the Institute of Indian Studies (Collège de France). AyurYog. http://www.ayuryog.org/blog/arion-roşu-collection-institute-indian-studies-collège-de-france.

Bendixen, Adelheid. 1990. "Überlegungen zu Rasaratnasamuccaya (Kapitel 1-11) und zu Rasendracūḍāmaṇi." *Journal of the European Āyurvedic Society* 1: 165–166.

Dole, V. A. 2015. *Srī Vāgbhatāchārya's Rasaratnasamucchaya*. Varanasi: Chowkhamba Press.

Hellwig, Oliver. 2009. *Wörterbuch der mittelalterlichen indischen Alchemie*. Groningen: Barkhuis.

Maas, Philipp. 2017. "Rasāyana in Classical Yoga and Ayurveda." *History of Science in South Asia* 5 (2): 66–84.

Meulenbeld, G. Jan. 2000a. *A History of Indian Medical Literature*, Volume IIA: *Text*. Groningen: Egbert Forsten.

Meulenbeld, G. Jan. 2000b. *A History of Indian Medical Literature*, Volume IIB: *Annotation*. Groningen: Egbert Forsten.

Miśra, Siddhinandan. 2009. *Rasendracūḍāmaṇi of Somadeva*. Benares: Chaukhamba Orientalia.

Mookerji, Bhudeb. 1927. *Rasa-Jala-Nidhi or Ocean of Indian Chemistry and Alchemy*. Volume 1. Calcutta: Nababibhakar Press.

Nath, Geethu S. 2017. "Concept and description of Rasaśāla in Rasaratnasamuccaya." *International Journal of Sanskrit Research* 3 (6): 195–197.

Radhakrishna Shastri. 1952. *Ānandakanda*. Volume 1. Srirangam: Sri Vilasam Press.

Rasa-Jala-Nidhi. See Mookerji 1927.

Rasaratnasamuccaya of Vāgbhaṭa. See Bapta 1890; Dole 2015.

Rasārṇava. See Ray 1910.

Rasaratnākara of Nitya Nātha Siddha. See Āchārya 1939.

Rasendracūḍāmaṇi of Somadeva. See Miśra 2009; Roşu unpublished.

Ray Praphulla Chandra Ray. 1910. *Rasārṇava*. Calcutta: Hitavrata Chattopadhyaya.

Roşu, Arion. 1997. "À Propos de Rapports Entre Rasaśāstra et Tantra: Étude Sur un Fragment du Rasendracūḍāmaṇi." In *India and Beyond: Aspects of Literature, Meaning, Ritual and Thought: Essays in Honour of Frits Staal*, edited by Dick van der Meij, 408–438. London: Routledge.

Roşu, Arion. unpublished. *Rasendracūḍāmaṇi (Le joyau de diadèmè pour mercure). Traité sanskrit d'iatrochimie de Somadeva. Texte, établi, traduit et commenté par Arion Roşu avec le concours de l'Université Hindoue de Bénarès.*

White, David Gordon. 1996. *The Alchemical Body*. Chicago: University of Chicago Press.

Map 5.1 South Asian manuscript records for the *Rasaratnākara*.

5

Alchemical Pilgrimage

Chapter 8 of the *Rasaratnākara*'s *Rasāyanakhaṇḍa*

Keith Edward Cantú

The excerpt analyzed in this chapter is the "Eighth Instruction" (*upadeśa*) of the *Rasāyanakhaṇḍa*, a section of the circa fifteenth-century[1] CE *Rasaratnākara* that is attributed to one Nityanātha Siddha, the son of Parvatī, and not the alchemist Nāgārjuna as several previous scholars had thought.[2] It describes what actions an adept (*sādhaka*) should take in order to benefit from the alchemical wonders of Srisailam (aka Śrīśaila, in this text also called Śrīparvata and Śrīgiri), the "Holy Mountain," an important historical site situated on the border between the contemporary Indian states of Andhra Pradesh and Telangana. The Eighth Instruction translated here is not a treatise on laboratory alchemy but at once a pilgrimage guide and a fantastic story, taking the listener—or, in our contemporary case, the reader—through a veritable wonderland of magical stones, plants, gods and goddesses, alchemical reactions, metals, mantras, creatures, and even treasure-hunting. It is, of course, expected that would-be adepts need to bring their shovel

[1] This is the date that Meulenbeld proposes, due to references to both Gorakṣanātha (as earlier pointed out by David White) and the *Kākacaṇḍeśvarīmata*, as well as the fact that the author Nityanātha was known to alchemical authors in the sixteenth century; for more on this question of dating see Meulenbeld 2000: 664–665.

[2] For the question of attribution see Dominik Wujastyk 1984: 75–78 and Wujastyk et al. 2022. Wujastyk has conclusively demonstrated that the Bengali historian of Indic science Praphulla Candra Rāy (1861–1944), followed by Mircea Eliade (1907–1986) and other scholars, had originally attributed the *Rasaratnākara* to the semi-mythical alchemist Nāgārjuna. Wujastyk proved that this is a false attribution for primarily two reasons: (1) Ray's misreading of manuscript excerpts of the *Mantrakhaṇḍa* as well as a mislabeled manuscript (MS Jammu Raghunatha 3153) of the *Rasendramaṅgala* with mixed leaves from the *Rasaratnākara*, and (2) the *Mantrakhaṇḍa* quotes portions of a text on Indian magic called the *Kakṣapūṭa* that is in fact attributed to an alchemist named Nāgārjuna, and it appears that Rāy accordingly ascribed the authorship of the whole *Rasaratnākara* to Nāgārjuna. The alchemist Nāgārjuna is also associated with Srisailam in the *Rasendramaṅgala*; see Wujastyk 1984: 76–77 and David Gordon White 1996: 164.

Keith Edward Cantú, *Alchemical Pilgrimage* In: *Indian Alchemy*. Edited by: Dagmar Wujastyk, Oxford University Press.
 DOI: 10.1093/9780197805831.003.0006

Figure 5.1 An engraving of what appears to be a Siddha, behind which is placed the sun and moon, from the enclosure (*prākāra*) of the Temple of Mallikarjuna in Srisailam. Photo © Keith Cantú.

The very first verse's mention of the god Mallikārjuna and other textual evidence allows us to conclusively pinpoint the text's geographical descriptions, not just to any "Holy Mountain" but to the physical environs of Srisailam itself. Many of the sites mentioned, such as the Plantain Forest (*kadalīvanam*), Chandragupta's Fort, and Alampur (today in the Jogulamba Gadwal district of Telangana), can still be located on contemporary maps, although their precise areas may have shifted over time. Furthermore, our text's division of the site into various "stations" or, more literally, "gates" (*dvāra*) that lead in the cardinal and diagonal directions of Srisailam remains a salient feature of the pilgrimage site even today. If the Eighth Instruction was not composed in Srisailam, it was composed by someone who was intimately familiar with the lay of the land.

The twentieth century saw a renewed interest in English-language scholarship on Srisailam, much of which should be read alongside this chapter in order to get a fuller picture of the site's significance. The scholar Arion Roşu (1924–2007) supplied what is probably the most detailed and succinct piece of academic writing on Srisailam to date, synthesizing a large amount of information that he gleaned from texts, including not only Hindu, Jain, and Buddhist references but also European medieval and colonial-era histories and an Old Russian travelogue that may refer to the site.[3] The same year saw another publication by M. Rama Rao (1906–1975), whose work *Temples of Srisailam* supplied the site with a rich foundation of relevance to archaeology and art history and included many photographs as well.[4] Rao's publication three years prior on temples built during the medieval Kākatīya dynasty also included a relevant portion on the temple to Tripurāntakeśvar, today located in modern Tripuranthakam, Andhra Pradesh, about ninety-five kilometers to the east of Srisailam.[5] This is a location on which our Eighth Instruction devotes almost twenty verses (14–33.1), and its presence helps contextualize the span of geographical space that the text is dealing with. These works were later followed by a detailed study by the late archaeologist P. V. Parabrahma Sastry (1920–2016) on the history of Srisailam. Sastry's contribution is especially useful for its records of extant Sanskrit and Telugu epigraphical inscriptions, all of which add more depth to the many references to proper names and places in our text's narrative.[6] An important and groundbreaking study on Srisailam was published by Bruno Dagens in French.[7]

When the earlier works of Rao, Roşu, and Sastry are placed in conversation with contemporary scholarship, a lively portrait emerges of a site with multifaceted social, religious, and even artistic[8] dimensions, especially for Śaiva sects such as the Pāśupatas[9] and, as Elaine Fisher has demonstrated, for the Vīramāheśvaras.[10] As Bisschop has revealed, one recension of the *Skandapurāṇa*, a text that also includes references to Pāśupata yogic practices, even contains a reference to Srisailam under its alternate name Śrīparvata. This is found in a portion of the text on Śaiva sacred sites that dates to around the sixth century CE. "There is a mountain there [in Andhra], the dwelling of Śrī, called Śrīparvata, a supreme field of success (*siddhikṣetra*) with hundreds of Siddhas and immortals. Śilāda installed a thousand auspicious liṅgas of Śūlin there. Human beings who see Īśāna, who

[3] Roşu 1969: 23–58. This article was significantly truncated for its English translation, which appeared over two decades later (Roşu 1992). I am grateful to David White for sharing his knowledge of these articles with me.

[4] Rao 1969.

[5] Rao 1966: 41–48.

[6] Sastry 1985.

[7] Dagens 1984.

[8] For a view of Srisailam from an art historical perspective see Shaw 1997.

[9] Sastry 1985:18 opines that most of "the Siva temples in Andhra were in those days [twelfth to thirteenth centuries] controlled by this Paasupata Saiva," who "used to command religious influence among the rulers and their subordinates of Andhradesa." For a detailed overview of Pāśupata literature, especially prior to this period, see Sanderson 2014.

[10] Fisher 2019.

dwells on Śrīparvata, are freed from the fear of birth and death. Brahmins who practice supreme yoga there attain the siddhi longed for in this very body. Even they who visit Śrīgiri only in their mind do not go to Yama's abode when they die."[11] The site's rich history also continues to inspire contemporary pilgrimage today, as Prabhavati Reddy has analyzed.[12]

The degree to which this site intersects with traditions of Indian alchemy[13] and haṭhayoga[14] has been a topic of interest in previous decades up to the present as well. At the same time, there is nevertheless a troubling lack of physical evidence for any practice of mercurial (laboratory) alchemy at Srisailam.[15] In the absence of such physical evidence, therefore, our text forms a critical record of these connections, since several verses allude to not only to the alchemical *saṃskāra* of "transmutation" (*vedha*) but also "goal-oriented practices" (*sādhanas*, e.g., mantra recitation and meditation) and powers (*siddhis*, e.g., *khecaratva* or "flight"[16]), all of which appear to reinforce the conception that, at least for some audiences, there was some degree of cross-pollination—if only textual—between these domains.

The Place of the Eighth Instruction in the *Rasāyanakhaṇḍa*

As mentioned above, the *Rasāyanakhaṇḍa* of Nityanātha Siddha is commonly held to be the fourth section (*khaṇḍa*) of the alchemical text *Rasaratnākara*, "The Jewel Mine of Mercury," and ostensibly treats on *rasāyana* or the "elixir," or "elixir regimen."[17] The other seven chapters or "Instructions" of the *Rasāyanakhaṇḍa* treat subjects as varied as killing

[11] Bisschop 2006: 68. I am grateful to Dagmar Wujastyk for bringing the relevance of Bisschop's study to my attention.

[12] Reddy 2014.

[13] White 1996.

[14] Mallinson 2019.

[15] In the words of David White: "What is most troubling is the fact that there exists no hard evidence—geological, chemical, archaeological, or epigraphical—to indicate that mercurial alchemy was ever practiced at Srisailam! Inasmuch, however, as it was a hub of Śaiva activity, prior to and following the advent of the Vīraśaivas . . . Srisailam was in fact linked to those regions, to the west and north, in which both alchemical raw materials and expertise were abundant" (White 1996: 112).

[16] For a brief but helpful summary of scholarship on the power of flight in Sanskrit, see Mallinson 2007: 183n113. For a more general overview of the *siddhis* in the context of Śaivism see Vasudeva 2012.

[17] For a general overview of *rasāyana* as an elixir in Indian alchemy and medicine, or as the regimen in which the elixir is applied, see Wujastyk, Newcombe, and Barois 2017 and Wujastyk 2017 in that same special issue. For its relatively rare usage in *haṭha-* and *rājayoga* literature, see Birch 2018.

mercury (*rasamāraṇa*), perfecting the body (*dehasādhana*), elixirs (*rasāyana*), shampooing and dyeing one's hair (*udvartanakeśarañjana*), and reinforcing potency and enlarging the penis (*vīryastambhanaliṅgav ardhana*).[18] The other four sections (*khaṇḍa*s) of the *Rasaratnākara* are "Mercury" (*Rasa*), "Lord of Mercury" (*Rasendra*), "Theory" (*Vāda*), and "Mantra" (alternatively called *Siddhakhaṇḍa*).[19] A philological study of all extent manuscripts of the *Rasaratnākara* remains a desideratum for extended research into this text,[20] especially since not all manuscripts appear to include the *Rasāyanakhaṇḍa*. Any statements about their dates, intertextuality, and area of composition must therefore still be tentative.[21] Even focusing on just the *Rasāyanakhaṇḍa* alone reveals some degree of variation in the contents of the sections. For example, one manuscript exemplar of the *Rasāyanakhaṇḍa* (MS 5-3091 of the National Archives, Kathmandu) that was copied at the Kasthamandap at Kathmandu includes the previous seven Instructions but curiously omits the Eighth Instruction.[22] At the same time, the two manuscripts that Jâdavjî Tricumjî Âchârya consulted for his 1913 edition, one belonging to Vaidya Dāmodara Viṭṭhala Damaṇakara of Bombay and another to Amṛta Vināyaka Jāmbekar of Barodara, must have included the Eighth Instruction since a few variant readings are cited throughout the text.

In any event, the Eighth Instruction does participate in the *Rasāyanakhaṇḍa*'s wider focus on mercury and the transmutation of metals in the context of perfecting, enhancing, and rejuvenating the body through ritual actions (*dehasādhana*, see especially the Second Instruction) and the consumption of prepared substances, including sexual fluids.[23] The application in the Eighth Instruction is certainly different, however, in that laboratory alchemy is minimized, and rites, mantras, and the magical

[18] This is description of the chapters derives from a helpful summary of contents provided in Wujastyk 1984: 73.

[19] For more detailed descriptions of these sections and their contents see Wujastyk 1984; Meulenbeld 2000: 653–665; and Wujastyk 2017: 14n40.

[20] Madhusudan Rimal at the University of Alberta has produced a preliminary and currently unpublished description of the digitized *Rasaratnākara* manuscripts of the National Archive of Kathmandu.

[21] See Wujastyk 2017: 13–14 for a summary of these issues as present the alchemical corpus. Here she also cites (14n30) Hellwig's assertion that the Vāda section of the *Rasaratnākara* may be among the earliest strata of alchemical works.

[22] In place of the Eighth Instruction, we find a colophon that provides details about its place and time of composition.

[23] Meulenbeld 2000: 663 notes that the Third Instruction of the *Rasāyanakhaṇḍa* (3.197–199) mentions a ritual in connection with a substance called *kālāntarasa* that "prescribes sexual intercourse with a menstruating woman and the collection of the mixture of male and female procreatory fluids (*rajobīja*) that is called *gagana*."

efficacy of deities, stones, herbs, rituals, mantras, and so forth are brought to the fore. Despite this key difference, it is not a clear dividing line since the *Rasaratnākara* more broadly does participate in this world. For example, the First Instruction of the *Rasakhaṇḍa* emphasizes the importance of an "Aghora Mantra" in the context of the worship (*pūjana*) of the alchemical process (*saṃskāra*) of mercury, a mantra that is alluded to again in the Eighth Instruction (verse 184). The First Instruction of the *Rasāyanakhaṇḍa* also places importance on intermediary deities, such as the Kumārīs and Yoginīs, and the text as a whole contains a large number of mantras.[24]

The "Mythology of Tantric Alchemy"

Despite this different emphasis, I would argue that the Eighth Instruction of the *Rasāyanakhaṇḍa* remains valuable as an example of the diversity of Indian alchemical literature, some of which extends beyond laboratory instructions into what David White has called the "mythology of tantric alchemy."[25] These wonders of Srisailam as expressed in the *Rasāyanakhaṇḍa* continued to have staying power in the alchemical corpus, as well. As several scholars have noted, a modified form of our instruction found its way into the *Ānandakanda*, "Bulb of Bliss," an encyclopedic compendium on alchemy that has been translated into Tamil and has stayed in print up to the present day. As for the Sanskrit syntax of the Eighth Instruction, the language—like the rest of the *Rasāyanakhaṇḍa*—mimics the style of alchemical manuals with its heavy use of prescriptive commands in the optative verbal tense, which is of course a style also utilized not just in ayurvedic texts but also in Buddhist and Hindu tantric literature. From a literary perspective, the whole instruction chronicles the way in which authors of alchemical mythology were able to find an established place for their literature—the substance of which is curiously resonant with today's fantasy literature and role-playing games[26]—by adapting the model of optative prescriptions that could be found in extant manuals of other genres, such as tantric and medical literature. As a result, I find that the creative vision of the author (or authors, if Nityanātha was a pseudonym) is discernible not just in

[24] Meulenbeld 2000: 663–664.

[25] White 1996: 74.

[26] This author while translating couldn't help but notice weird similarities between some of the adept's tasks in this Eighth Instruction and certain questlines in, say, The Legend of Zelda or The Elder Scrolls series.

the excruciatingly specific instructions but also in the rewards and treasures, which are truly the stuff of legend. In this respect, the text could also be situated in a wider genre of Śaiva tantric literature that refers to so-called middle powers (*madhyamasiddhis*), some of which also include powers to perceive treasures hidden under the ground.[27]

Who, then, would be fit to undertake the rites of Srisailam, if not a laboratory alchemist? The unnamed traveling "adept" (*sādhaka*, related to the more common derivative "sadhu") is of course also a staple figure in both Buddhist and Hindu yoga and Tantra, but in the context of the Eighth Instruction certain other technical knowledge is assumed.[28] He is given various other titles according to his proficiency at carrying out various tasks, including "hero" (*vīra*) and "accomplished one" (*siddha*). From the text we can infer that he is to engage in each "goal-oriented rite" (*sādhana*) with the proper mantras, he is to be unwavering (*nirvikalpa*), and he is to be conversant in herbalism, metallurgy, and astrological time (especially the lunar *tithis*). He is also expected to be quite a versatile worker, as at numerous points he is expected to know how to tie garments to bamboo poles, carry around a child, and even deliver food to a hungry "nymph" (*yakṣinī*).

All things considered, the fantastic nature of our text's instructions appears intended, on the one hand, to fulfil religious functions specific to Srisailam rather than serve as a manual for laboratory alchemy per se. These functions would likely include the encouragement of pilgrimage to the region's many diverse areas as facilitated by the stimulation, whether consciously or not, of the religious imagination among would-be devotees (*bhaktas*) and adepts (*sādhakas*) who may have been already somewhat familiar with the mythology and were predisposed to taking the journey. On the other hand, I think it would be over-cynical to discount the possibility that there very likely were individual adepts who believed these instructions would work and therefore would be interested in pursuing such a difficult quest for treasure and bodily immortality. What survives therefore is a fascinating record of both social and individual relevance, that is, of not only the power ascribed to a temple area by authority but also a reflection of the desires of individuals who sought to benefit from a symbiotic relationship with its natural environment.

[27] Vasudeva 2012: 271–276.

[28] The adept in our text is always declined in the masculine gender, but, at the same time, the text never explicitly denies the potential for feminine adepts (*sādhikās*).

Notes on the Text and Its Translation

The translation in this chapter follows the edition of the *Rasāyanakhaṇḍa* prepared by Vaidya Yādavjī Trikamjī Ācārya, first published in 1913 and reprinted in 1939 and with the "*Rasacandrikā*" Hindi commentary by Indradev Tripāṭhī in 1982. Yādavjī Trikamjī consulted at least two manuscript variants for the Eighth Instruction (see above), both of which at that time were in private collections and could not be tracked down for this translation. However, I was able to locate the Eighth Instruction in a copy of one manuscript source of the *Rasāyanakhaṇḍa*: R. No. 317 GON, held at Gujarat Ayurved University, Jamnagar, which I have consulted alongside the above published sources.[29] Whenever possible I have compared my translation to the manuscript and cross-checked the alchemical terminology of metals and various processes with the list of helpful definitions given in the appendix to the *Rasārṇava*, a text that was also published with a "*Rasacandrikā*" Hindi commentary by this same scholar.[30] I have also consulted the parallel version in *Ānandakanda* 1.12 in numerous instances to clarify the meaning of the text. Finally, I am grateful to Dagmar Wujastyk and Patricia Sauthoff for their helpful comments while reading portions of the text together, as well as to Oliver Hellwig for his Digital Corpus of Sanskrit that proved very helpful.

I have attempted to translate proper names into English equivalents whenever possible to give the reader a sense of the aesthetic context of the text. I have also done this for the units of weights and measurement, rounding to the nearest whole number of a viable metric equivalent, and for determining this equivalent I have mostly followed the correlations supplied in Tripāṭhī's Hindi commentary. It has not been possible for me to determine if these correlations with contemporary measurements are exact since there are variations in definitions across different genres and eras of Sanskrit literature; Tripāṭhī himself supplies no less than six tables of weights (including the contemporary metric system). In any case, such correlations should at least provide the reader with a clearer understanding of the general range of weight or measurement at play with the various substances and geographical references. Here are two conversion tables that I have constructed for ease of reference:

[29] I have not included here the many extant manuscripts of the *Rasaratnākara* in which the *Rasāyanakhaṇḍa* is absent (e.g., London Wellcome 107920).

[30] Tripāṭhī and Tāradattapantaḥ 1978. I am grateful to David White for letting me borrow his personal copy of this source.

Measures Occurring in the Eighth Instruction

a fingerbreadth	1 *aṅgula*	—	~ 1.763 cm (0.69 in)[31]
a forearm	1 *hasta*	24 *aṅgulas*	~ 42 cm (16.5 in)
the space of a bow	1 *dhanvantara* / *dhanu*	4 *hastas* / 48 *aṅgulas*	~ 168 cm (5.5 ft)
a person (with arms outstretched)	1 *puruṣa*	120 *aṅgulas*	~ 2 m (6.6 ft)
a calling-range; Indian league	1 *krośa*	2,000 *dhanvantaras* / 8,000 *hastas*	~ 3 ½ km (2.2 mi)
a range for grazing	1 *gavyūti*	2 *krośas*	~ 7 km (4.3 mi)
a distance traveled without unyoking	1 *yojana*	4 *krośas*	~ 14 km (8.7 mi)

Weights Occurring in the Eighth Instruction

a rosary pea	1 *guñjā*	—	~ 120 milligrams
a coin	1 *niṣka*	3 & 1/3 *guñjās*	~ 4 grams
a weight	1 *karṣa*	3 *niṣkas*	~ 12 grams
a fluid measure	1 *pala*[32]	4 *karṣas*	~ 48 grams
a measure of capacity	1 *prastha*	—	~ 81 grams

Abbreviations

ĀK	Sastri, S V Radhakrishna, 1952
Hi.	Hindi
GAU	*Rasaratnākara (Caturtha: Rasāyanakhaṇḍa)*, R. No. 317 GON of Gujarat Ayurved University, Jamnagar
JTĀ	Âchârya, Jadavjî Tricumjî 1913
JTĀ (kha)	Variant reading given in JTĀ
MW	Monier-Williams, 1899
RĀ	Tripāṭhī Ayurvedācāryaḥ, Indradev, 1978, 353—442
RC	Ācāryaḥ, Yādavjī Trikamjī 1982

[31] For this number I have followed Balasubramaniam 2009: 681–682. Although the Maurya Period predates our text by at least a millennium, it provides a general rule of thumb (pun intended) for this particular measure.

[32] A *pala* can also be a liquid measurement. It is unclear which it refers to in the one instance in which *pala* occurs in the Eighth Instruction.

Rasaratnākara Rasāyanakhaṇḍa, Chapter 8

1. The body's powers come forth quickly by means of trees, soil, bulbs, and waters in the Holy Mountain (Śrīśaila). This treatise is just the essence of what was spoken of and beheld by Śambhu in its very dense entirety.
   Having spoken using many arguments about that evident accomplishment that is the cause of all happiness, I will now speak of that going on the adepts' path of experience for the sake of enjoyment and liberation.
2. A stone resembling an elephant stands in front of the White-Jasmine God (Mallikārjuna) that always flows with guggulu at night.
3. [The adept] should put what has been harvested into a bottle-gourd bowl and the same amount of sulfur (*gandha*) together with flame-of-the-forest wood (*pālāśa*). One should always eat four grams of it.
4. These remove old age and death after a month, and one will live as long as there are moon and stars.[33] He should add this guggulu by the tens of millions into melted copper (*tāmra*).

5ab. It will become divine gold (*kanaka*) in equal splendor to that which comes from the Jambu River.

5cd. The Accomplished Lord of the Bell (Ghaṇṭāsiddheśvara) is situated to the left of the Jasmine Lord (Mallinātha, i.e., Mallikārjuna).

6–7. He sees a pool at this gate, and a bell hangs there. At night during the dark lunar fortnight, a rite is to be performed there by three people who have fasted, who are unwavering, and are awake. One should bathe the god, and another should carry the water.

8. The other should sound the bell continuously until the night. Śambhu ought to then be pleased and bestow the power of flight.

9. He should dig at a distance of three and a half kilometers[34] to the south of The Accomplished Lord of the Bell. Some soil will be seen there that resembles an ox bezoar (*gorocana*).[35]

10ab. He should drink twelve grams[36] of this with white sugar (*sitā*) and milk, and he will become immortal after a week.

[33] *ācandra-tārakam*. This is a recurrent promise in this and related alchemical texts.

[34] = *krośamātrake*] GON, JTĀ. *jānumatrakam*, "as far as the knees" JTĀ (kha).

[35] This is probably a cow gallstone, which is a bright yellow-orange color.

[36] = a *karṣa*.

10cd. As for the celebrated Moon Water (Candrodakaṃ), it is to the west of the Jasmine Lord.

11. It remains there in the month of Vaiśākha.[37] The adept, staying close to it for the whole month, should properly perform the rite without wavering.

12–13. He should perform mantra-recitation at night. At midnight its water is to be touched. When the moon goes out, then, quickly making a gesture of salutation with the hands (*karāñjali*), he should stand there. When its water comes, he should drink it at that moment. He shall become a hero with an adamantine body and shall live as long as there are moon and stars.[38]

14ab. The god who is Śiva as the Destroyer of the Three Cities (Tripurāntaka) is at the eastern gate of the Holy Mountain.[39]

14cd. There is a tamarind tree (*ciñcā*) to the north of the god.

15. It is known that Bhairava himself will be seen at its roots. He should dig into the earth two meters[40] in front of it.

16. A heated pool with blue-colored water will also be seen. He should bind the leaves of the tamarind tree with the proper cloths.

17–18. They will become fish after placing them in the pool there. Having collected these, he should cook them with its wood. The adept ought to leave behind their head, their tail, and their bones, and he should eat the remainder. He will become faint for an instant. By this means he will arise as one who is himself awakened.

19ab. The man sees the fertile soil split,[41] and ought to live ten thousand years.

[37] This is a month between April and May in the Gregorian calendar.

[38] This passage presents some translation difficulty due to there being three possible subjects construing with various verbs: (1) the adept reciting the mantra, (2) the moon, and (3) the water. As a result, several alternative translations are possible.

[39] This refers to Śiva as one who "makes an end" to the cities of *asura*s or enemies of the gods. For more on Tripurāntaka see Rao 1966: 41–48.

[40] = a *puruṣa*. According to Patrick Olivelle, referencing the *Kātyāyana Śrautasūtra* 16.8.22, the length of a *puruṣa* or "height of a man" dates back to Vedic texts where it generally measured 120 *aṅgulas*; see Olivelle 2013: 554n2.20.19.

[41] According to David White (personal correspondence), this may refer to a power of perceiving "cracks" or "fissures" in the earth, visible to persons with a kind of x-ray-like vision, which are portals to hidden treasure. If this is true, it may indicate a kind of *khanyasiddhi* or "power to find treasure troves" (Vasudeva 2012: 269). However, the text itself unfortunately gives no indication of what *chidraṃ paśyati medinyāṃ* ("he sees a split in the earth") exactly refers to.

19cd–20. There is a village named "Settlement of the Jewel" (Maṇipalli) fourteen kilometers[42] to the west of the god Śiva as the Destroyer of the Three Cities. A mountain is to its west. On its western side an auspicious door will be seen.

21–22. He should enter from the east, and then should go to the south. Fine stones with a blazing radiance, in the shape of mangos (*āmra*), will be seen for as far as seventeen meters.[43] These are to be grabbed. He should bind them with a cloth. The cloth ought to become blood-red, and then he should place this in milk.

23. This milk will turn blood-red, and the adept should drink this milk. After a week, perhaps, he will have an adamantine body, and will live as long as there are moon, sun, and stars.[44]

24–25. The Female Cuckoo's Cave (Kokilābila), famous among all people, is to the north of the god who is the Destroyer of the Three Cities (Tripurāntaka). Having undergone emesis and purgation, the adept should enter there as far as seventeen meters.[45] Stones in the shape of cuckoos stand there, and he should fetch them.

26–27. Sesame seeds (*tila*) that are thrown onto them instantly burst. Having thus seen this proof and rubbing them together, he should cast them into the milk. The milk turns black.[46] He should drink this until satiated.[47] The man will become like a divine poet-sage.

28. He will live as long as the three days of Brahma, free of wrinkles and gray hair. He, being endowed with great strength and as quick as the wind, sees the fertile soil split.

29–30. He will see a god named "One who Possesses the Heavenly City" (Svargapurī) within thirteen kilometers[48] to the east of the god who is the Destroyer of the Three Cities. There in front of the

42 = two *gavyūtis*.

43 = ten *dhanvantaras*.

44 This and several subsequent verses bear a striking resemblance to the *Nidhipradīpa* of Siddha Śrīkaṇṭhaśambhu, of which the dating is unknown. According to Vasudeva (2012: 272–275), the text mentions an "occult science of *paṭṭas*" (here likely strips of cloth), or prognostication based on iconic images that may appear when a cloth is smeared with various concoctions. Another chapter in that work apparently gives recipes for various kinds of occult collyrium and eye-salves that make it possible to find treasure hidden in the ground, which is a recurring theme in this "Eighth Instruction."

45 = ten *dhanvantaras*.

46 *kṛṣṇa*. Perhaps the milk turns blue or dark, rather than black.

47 *ākaṇṭhāntam*, "up to the end of his neck."

48 = *yojana*.

god he should dig in the earth up to the length of the knees. There stones like fruits appear, yet they are sharp to the touch.

31–32. A door to a cave will be found seven kilometers[49] to the west of the god who is the Destroyer of the Three Cities. He should go inside for five meters.[50] A black-colored date palm (*kharjūravṛkṣa*)[51] will be seen there. Having drunk the juice of its fruits, stupor will arise for an instant.

33ab. He will arise as an accomplished one (Siddha), and ought to live as long as there are moon, sun, and stars.

33cd. The famous gods Lord of the Web (Jāleśvara) and Lord of Heroes (Sureśvara) are found at the southern gate of the Holy Mountain.

34. He should dig into the earth there. Stones come forth in the shape of auspicious fruits that break [other materials] upon touching.[52]

35. The god Lord Rāma (Rāmeśvara), is found there in its vicinity. Stones are there in the shape of rudrakshas, and they are sharp to the touch.

36ab. He stands on one foot in the banyan tree (*vaṭa*) that confers radiance.

36cd. The god named the Whirling One (Āvarta) is adjacent to it in the west.

37. A mountain will be found. Stones resembling orpiment (*tālaka*) are there. Having taken them, he should intensely blow on them. Divine gold shall appear.

38. The Lord of the Coils (Kundaleśvara) will be there at the southern gate. He should dig adjacent to him as far as the knees and carry the dirt away.

39ab. Stones, blood-red and in the shape of auspicious fruits, are there that cause transmutation[53] when touched.

[49] = half a *yojana*.

[50] = *dhanvantara*

[51] *kharjūravṛkṣa*] GON, JTĀ. *arjunavṛkṣa*, "arjuna tree" JTĀ (kha).

[52] *sparśabhedaka*. The translation follows RC's reading (Hi. *sparśamātra se pūṭne vāle patthar hāĩ*).

[53] *sparśavedhaka*, "that which pierces upon touching" or "that which penetrates upon touching." This is a reference to the stones' power of transmuting base metals to gold or other precious metals, similar to the legendary "philosopher's stone" in western alchemy. In verse 125 below this is made explicit by the author's inclusion of the phrase "of all metals" (*sarvalohānāṃ*). See White 1996: 265–269 for the role of *vedhana* as alchemical transmutation in the context of the alchemical *saṃskāras*. Also see Dagmar Wujastyk on the *saṃskāras* in the *Rasahṛdayatantra* in this volume.

39cd. The pool of the god Lord of Man (Puruṣeśvara) is nearby.

40. Two trees also stand there, rosary pea (*guñjā*)[54] and soapnut (*riṣṭaka*). He will instantly become immortal after consuming their leaves.

41–42. The Elephant Stone (Hastiśilā) is also there. To the south of it he should dig about a forearm's length.[55] Then he should look. The aforementioned stone there, in the shape of a plum (*jambūphala*),[56] is to be carefully handled. Once it is known by the name of Elephant Head (Hastiśiraḥ) then it indeed becomes the Elephant Stone.

43–44. It is said by Śiva that the celebrated Shady Parasol (Chāyāchattra), in the form of Śiva, stands fourteen kilometers[57] to its south. It measures a hundred forearms and encircles that place. One who goes under it will become invisible to everyone.

45. As a Siddha he shall comparable to Rudra, who plays in the three worlds. Or he shall attain the eight powers that are famous far and wide.

46. Magnetic iron (*kānta*), mica (*abhra*), and gold (*kañcana*) are to be rubbed with mercury (*sūta*), which becomes superior by degrees. He should gradually and with effort cause this ball of mercury to be enveloped by these three metals.

47–48. Having bound [the ball] in a garment, he ought to firmly fasten this garment to a bamboo shoot (*vaṃśa*). He ought to invoke using Kālī's mantra and fasten this bamboo shoot to the Shady Parasol. The pill ought to be produced in a single instant. The adept ought to hold in his mouth that which grants the power of flight.

49. Having bound orpiment (*tālaka*) and red arsenic (*kuṇaṭī*),[58] he should fasten it to this bamboo shoot. He who has eyes smeared with these two sees the treasure which is extant on the earth.

50. Having bound a sword[59] to the shoot, he should place that which was enchanted[60] by the aforementioned mantra. He should hold

[54] According to RĀ, this is the "seed of Abrus precatorius" (rosary pea). In this context it likely refers to the climber that produces the rosary pea.

[55] = a *hasta*.

[56] On *jambū* as plums (and not the rose apple), see Wujastyk 2004.

[57] = one *yojana*.

[58] This could alternatively refer to a kind of coriander. However, the juxtaposition with orpiment makes red arsenic more likely.

[59] *khaḍga*. This may refer to iron instead of a sword.

[60] *mantrita*, "acted on by a mantra."

this sword in his hand, and he will become the conqueror of the three worlds.

51. Having bound a bezoar (*gorocana*), he should there cause it to enter the hollow of the shoot. He ought to make a mark on the forehead using this [resulting substance]. Upon being seen he will bewilder the people.
52. After placing antimony-based collyrium (*srotoñjana*) and kohl (*kajjala*) there like before, he whose eyes are smeared [with this mixture] will be invisible even to the gods.
53. Having put in one's shoes according to this method, one should direct attention to the two feet. He instantly will go wherever he wishes.
54. Having likewise placed a blood-red garment there, the adept who wears it becomes invisible. He will be seen when releasing the garment in the same way.
55. Having prepared a ball of magnetic iron (*kānta*), mica (*abhra*), gold (*kañcana*), and mercury (*sūta*), he ought to wrap it at the end of the bamboo shoot while reciting a mantra. It will transmute [metals] upon touching them.
56. At the western gate of the Holy Mountain is a lord named Brahmeśvara. Some say you should touch his lingam. One who resembles Durgā is also there.
57. That which is a nine-fold staircase is likewise situated in the river there. It stands at the station of the god. He will observe that it has the color of a mung bean (*mudga*).[61]
58. To the southwest of Brahmeśvara lies the Tamarind Forest (Tintiṇīvanam). A pool is situated there in the forest. Caṇḍikā remains there.
59. On a single foot he should fetch a leaf under a tamarind[62] from her tree. Having bound it in a garment, he should cast it into the pool and bathe there.
60. At the end of the bath, all the fish are to be gathered in a bundle. He should cook them with [the tree's] wood but should leave out their throat, tail, and head.

[61] *mudgavarṇa*. RC interprets this as the mung bean (Hi. *mūg*), which is usually yellow or green in color.

[62] *ciñcā* can also mean the "tamarind tree" itself. This translation follows the reading in RC, which gives Hindi *imlī*, "tamarind."

61. Having arranged a portion for the god, a second portion should be for a guest. The adept, having eaten the third portion, will become faint for a moment.

62. In an instant he will arise as a Siddha. He sees the fertile soil split. He will go forth as one who is forever young, never dies, and who is never suppressed, even by the thirty divines.[63]

63–64ab. "Bhīma's Footprints" (Bhīmapāduka) is a village in the northern part of Alampur. In the vicinity of this village he should dig into the earth up to forty-two centimeters deep.[64] Stones are there that are curved like snakes. They are to be taken and touched.

64cd. The famous deity who is called the Lady of Yoga (Yogeśvarī) is in Alampur.

65–66. There is a beautiful cave in front of the goddess. Inside it, he should dig into the earth. Stones resembling frogs are to be taken. Having blended them with feces (*viṣṭhā*), he add these to tin (*vaṅga*).[65] The tin will become silver (*tāra*). He who is clear-sighted should consume these stones together with honey and butter.

67–68. There is a deity named Maheśa at the northern gate of the Holy Mountain. The Mango Tree of Bees (Bhramarāmra) is situated there. He should make its ripe fruits burst open. Having obtained proof that bees, endowed with life, have come forth, he should abandon these bees.

69. He should cause the fruit and its juice to be cooked and should drink this juice according to his wish. He will become quite faint from this and, upon waking up, he should then drink some more.

70–71ab. He should act in such a way for three weeks. He will become endowed with a diamond-body. Free from wrinkles and grey hair, he shall live as long as there are moon and stars. [Such] a

[63] *tridaśa*. MW states that this can refer to actually thirty-three intermediary deities, rounded up to thirty: twelve Ādityas, eight Vasus, eleven Rudras, and two Āśvins. It can also refer to the "divines" more generally, as may very well be the case in this verse.

[64] = a *hasta*. The text does not specify whether a measurement of depth or length is referred to. Therefore, the directions may state that he should dig in an area of up to one forearm wide.

[65] I have translated this metallic substance as "tin" based on RC's reading (Hi. *vaṅga*), but it may also refer to lead.

man will become a knower of all languages and will become as quick as the wind.

71cd–72ab. He ought to solidify forty-eight kilograms[66] of liquified tin (*vaṅga*) by means of this same mango fruit (*āmraphala*) which is free of bees. He shall surely obtain silver (*tāra*) by placing it [in the fruit]—this is the truth that the Lord (Īśvara) has spoken.

72cd–73. Having perforated this fruit, a bee will then be made to fly up. Having placed inside twelve grams[67] of pure mercury (*pārada*) and the same amount of black mica (*kṛṣṇābhraka*), he ought to close the opening and smear it with cow dung and soil.

74. He should cook this [mixture] that was dried in the shade for three hours along with the decoctions[68] that are produced from these mangoes.[69] The pill is cool by nature and is to be taken inside a fruit.

75–76ab. Placed in the mouth it grants flight, invisibility, and great power. By means of this decoction he shall become like the Master of Speech (Vākpati).[70] He shall live as long as there are moon and stars. He shall cause silver, copper, and lead (*bhujaṅga*)[71] to be transmuted by the tens of millions.

76cd–78. He should place black mica (*kṛṣṇābhra*) and pure mercury (*sūta*) into this fruit as before. He should likewise smear it with soil and cow dung, and ignite it using cakes of wild cow dung. This potion, cool by nature, is to be taken, and he should always eat an amount equal to a rosary pea (*guñjā*)[72] together with honey and ghee. After a month the man will become a young boy. The hero, endowed with the power of the Serpent-daemons (Nāgas), shall live as long as a day of Brahmā.

[66] = a thousand *palas*.

[67] = one *karṣa*.

[68] *kvātha*. This appears to be attested in GON (though the copy is faded) while JTĀ (kha) gives variant *kaṣṭha*, wood, which may indicate the wood of the mango tree.

[69] = for a *yāma*, which in this case indicates a period of time. RC gives "*ek pahar*," which can be a period of three hours or one-fourth of a day.

[70] RC understands Vākpati in astrological terms as Bṛhaspati, "Jupiter."

[71] The translation of *bhujaṅga* as "lead" instead of "snake" is based on RC's reading (Hindi *sīsā*). It is a synonym for the term commonly used for lead in alchemical and medical works: "*nāga*."

[72] *guñjā* "rosary pea" is also a standard measurement. RC correlates it to Hindi *rattī* and equates it to 120 milligrams.

79. He will be seen as famous among the people. Worship is to be performed by the adepts[73] using the non-conceptual mantra *oṃ huṃ phaṭ*.
80. After looking in the sky to the west, he should suddenly see divine flying chariots. Great trust will arise.[74]
81. Then he who is wise should head west and cross the river. A cavern[75] will be seen, and he should enter there at the western side.
82. He should go, alone and unwavering, for forty-two kilometers.[76] The Plantain Forest (Kadalīvana) will be seen, spread over seventy kilometers.[77]
83. Inside it there is an expansive lake, full of water, and within that is a lion-throne that looks like pure crystal (*sphaṭika*).

84–85. Having seen this, he should fall upon the earth like a pole, and should utter this mantra: "*oṃ hrīṃ vidyāvāgīśvarādhipataye hrīṃ oṃ svāhā*." After bathing he should make an effort to recite the mantra a hundred thousand times. After preparing a meal of bulbous root, roots, and fruits according to his wish, the Supreme Lord (Parameśvara) with four arms and three eyes, resembling pure crystal, will then be seen above the lion-throne.

86. After bowing with respect again and again with hymns, mantras, and adorations, there is no doubt that he will grant a boon to the adept, according to his wish.
87. There are hundreds of sacred places there. The number of wells is nine hundred, and there is the same amount of pleasure-groves and of forests called "Gladness" (Nandana).
88. Nine hundred pools will also be found in the Plantain Forest. There is same amount of cave entrances and even the same amount of wish-granting trees.

[73] ĀK gives the variant *sādhakaḥ*, a nominative singular form rather than the plural instrumental *sādhakaiḥ*. This could be an attempt to correct the text given that "adepts" in the plural are seldom referred to in the text.

[74] This is based on RC's reading (Hi. *isase acchī tarah viśvās ho jātā hai*).

[75] *gahvaram*. Alt. trans.: "woods," "thicket," "hiding place." The translation follows the reading in RC (Hi. *guhā*, "cave").

[76] = three *yojanas*.

[77] = five *yojanas*.

89. A nymph[78] named the Raveness (Maukalī) is also situated there in plain sight. She says, "Give me the food that I long for, if you wish."

90. Then he is to proclaim, "I will give it, O supreme Lady." He who is wise should cause rice cooked with milk or fruit and other foods[79] to be set before her.

91–92. She says, "This is my son. Hold him to your chest this very instant. As long as you do not let him fall onto the ground then the eight powers are yours. There is no doubt that I will serve you. I will wreak havoc[80] either after eating or after he is let go, so guard him with care." The adept who is a hero should act accordingly and shall obtain power.

93ab–94. He should go seven kilometers[81] to the north of that lake.[82] There is a Forest of Gladness called "Divine" (Divya), and it is filled with blossoms. He will not feel hungry or thirsty due to the blossoms' mere fragrance.[83]

95. He should nevertheless eat one fruit from there, according to his wish. The man shall be endowed with a diamond-body by means of eating this alone.

96–97ab. He should go for three and a half kilometers[84] to the south of that lake. A Forest of Gladness filled with pomegranate trees (*dāḍima*) will be seen there. He should eat this fruit. As a Siddha he shall live for a thousand ages.

97cd–98. He should go about two kilometers[85] to the east of that lake. A dark Forest of Gladness filled with amla fruit (*dhātrīphala*) is there. He should eat these fruits, and he shall live for three hundred aeons.[86]

99–100ab. He should go to the west of this lake for fourteen kilometers.[87] A Forest of Gladness filled with wood-apple

[78] *yakṣinī*: Alt. trans.: "tree spirit," "dryad."
[79] *phalāhara*: RC interprets this as "fruits, roots, and so on" (Hi. *phal-mūl ādi*).
[80] *karomi saṃhāram*: "I will cause destruction."
[81] = half a *yojana*.
[82] This is most likely the same lake mentioned in verse 83.
[83] "He will not feel" translates *na vidyate*: "hunger and thirst are not known."
[84] = one-fourth a *yojana*.
[85] = half a *krośa*.
[86] *kalpaśatatraya*. I have translated *kalpa* as "aeon" to give the Sanskrit non-specialist a sense of the magnitude of this amount of time, which varies according to the source.
[87] = one *yojana*.

fruit (*bilvaphala*) will be seen there. The hero should eat this fruit. He shall live as long as there are the moon, sun, and stars.

100cd–102. These are what are to be enjoyed in this Forest of Gladness. A lingam will be there, blue in color. He should enter it from the northern gate. A Serpent-daemon[88] with great strength will be seen there. He is ferocious, has the shape of seven serpents' hoods, and is exceedingly terrifying. He should perform an adoration for it, should perform "*hruṃ hruṃ*,"[89] and should get up. Then he says, "I will give you an invisibility chant."

103. [He also says:] "Go to the western gate. A divine maiden is there." The adept who is a great hero should go there, reciting the mantra.

104. She, being pleased, will grant you a necklace but she will not grant you entrance.[90] He ought to wear this necklace around his neck and he will manifestly become the Lord of Speech (Vāgīśvara).

105–106. But then he should go to south. There, upon seeing the terrifying Guardian of the Fields (Kṣetrapāla), naked and blue-colored with hair let loose, with a crooked gaze and armed with a mace, he ought to utter the mantra "*ha hā he he haṃ hruṃkāraṃ phaṭ huṃ*," with "*svāhā*."

107. The Guardian of the Fields settles down by means of the recitation of this mantra. He bestows the power of flight but never grants entrance.

108. Next, the Lord of the Troops (Gaṇeśvara) will be seen there to the east. The adept ought to worship him and then enter.

[88] *nāga*. According to MW (citing his *Religious Thought and Life in India* as well as Nāgānanda), Nāga are "of the race of Kadru or Su-rasā inhabiting the waters or the city Bhoga-vatī under the earth; they are supposed to have a human face with serpent-like lower extremities . . . their kings are Śeṣa, Vāsuki, and Takṣaka" (532). The understanding of *nāga* here may be more general, however, as a local spirit.

[89] It is unclear exactly whether *hruṃ hruṃ* is a mantra or some kind of ritual behavior. JTĀ notes a variant *huṃkāraṃ ca samuccaret*, "and he should pronounce the syllable *huṃ*" in place of "should perform *hruṃ hruṃ*."

[90] *na praveśāṃ*] GON, JTĀ. JTĀ (kha) notes the variant *prajeśatvaṃ*, which would alter the translation of the first part of verse 104: "She, being pleased, will grant you a necklace and will grant you lordship over creatures."

109. The divine bow[91] will be seen there, which is a fascinating lingam. He ought to worship it by means of Śiva's mantra, even the syllable "*ra*," and not otherwise.

110. Then, having fasted, he should recite mantras there day and night. Rudra will arise before his eyes. He will grant a boon according to what is desired.

111ab. These therefore, among others, are the powers of the Plantain Forest.

111cd–112. The Cave of the Lotus Woman (Padmāvatī), inscribed with letters, is situated fourteen kilometers[92] to the east of the lake. The adept should enter there, having undergone emesis and purgation.

113–114ab. He should go as far as about a hundred and seventy meters.[93] The hero, upon seeing a drum there, ought to strike it correctly. Then the Lotus Woman will come and give the nectar of immortality (*amṛta*), which will make him immortal upon drinking it.

114cd–116ab. Then she will yet ask of you, "Come to my temple and claim these divine maidens, amounting to five hundred, along with their entourage. I will become your wife as long as Brahmā shall live. When this is over there is no doubt that I will go to the supreme place."

116cd–117. The god who is Lord of the Kadambas (Kadambeśvara) will be found to the southeast. A kadamba (*kadambaka*) tree is situated in front of a pool to his east. Having collected its leaves, he should smear them with mustard oil (*kaṭutaila*).

118. He who is wise shall cook using the oil from the seeds of this tree or its wood. These all shall become fish. He should place them in a copper pot.

119. He should eat them with honey and ghee, leaving out the neck, tail, and head. That instant he will arise as a Siddha who is comparable to Rudra and who is endowed with great strength.

[91] *cāpa*. This follows RC's reading (Hi. *divya dhanuṣ*). The term could also mean an arc, or a rainbow.
[92] = one *yojana*.
[93] = one hundred *dhanvantaras*.

120. A tree will also be found at the location of the river near the Lord of the Coils (Kuṇḍaleśvara). Power will arise by means of performing the steps as before. There is no doubt in this matter.

121–122ab. The Mountain of Blossoms (Puṣpagiri) is situated to the north of the god who is Lord of the Pigeons (Kapoteśvara). He ought to circumambulate it deosil three times with great care. Then the man, having climbed to its peak, should receive the power of flight.

122cd–123. The Gate of the She-goat (Chelikādvāra)[94] is situated to its northeast. He should enter it from the north for five meters.[95] Stones in the shape of mice stand there. He should collect them.

124. Having ground up one of them, he should drink it with buttermilk (*takra*). He ought to fall into a swoon. After a moment he rises as a Siddha, and he shall live for three days of Brahmā.

125. He shall even become capable of transmuting all metals by touching them. He shall receive power after staying day and night before the god.

126–127. The Dual Deities (*devatādvayam*) are to the south of the god who is Lord of the Pigeons (Kapoteśvara). He ought to dig in between them, up to his hips. Stones resembling bezoars (*gorocana*) are there. Having gathered them, he should grind them together with honey and ghee. There is no doubt that, as a result drinking this, a mortal will become one who shall live for an aeon.

128–129ab. To the northwest[96] of the god who is Lord of the Pigeons he should dig as far as forty-two centimeters.[97] Stones resembling pigeons[98] are situated there that break when one touches them. He should take one while worshiping the deities.

[94] *chelikādvāra*] GON, JTĀ. JTĀ (kha) gives variant *ambikādvāra*, "the gate of [the goddess] Ambikā."

[95] = three *dhanus*.

[96] *vāyavye*: "in (the direction of) Vāyu . . . " RC adds "corner" (Hi. *koṇ*).

[97] = a *hasta*.

[98] *pāravata*. This follows the reading in RC (Hi. *kabūtar*).

129cd. Bhṛgu's Fall (Bhṛgupātana)[99] is to the northeast of the lord who is the White-Jasmine God.

130. A pool will be found in its vicinity. He should fetch soil from it. He should blow on it as a mixture with the five products of the cow (*pañcagavya*)[100] and with the charcoal (*āṅgāraka*) of acacia wood (*khadira*).

131ab–132ab. A metal resembling a blue lotus (*nīlotpala*) will now come upon it. He should make this to be sprinkled repeatedly with honey and ghee seven times. He should hold this pill in his mouth. The man will become comparable to Viṣṇu.

132cd–134. The famous mountain named Unshaded Oil (Acchatailā) is three and a half kilometers[101] in the eastern direction from Bhṛgu's Fall, at the opening of the ocean. Before it, to its west, is the door to a cave. When he has gone inside it for eight and a half meters,[102] a pool will be seen. Its mercury is resplendent like liquid lac (*lakṣārasa*).[103] After collecting it in a bottle-gourd (*alābu*) pot he will be able to transmute [metals] by the tens of millions.[104]

135. The Fort[105] of Candragupta,[106] famous among the people, is to the east of the god who is White-Jasmine. There a temple will be found.

136. A shrine (*caitya*) is situated to its east. To the east of the shrine is the famous, mung bean-colored Great Stone (Mahāśilā). It causes transmutation [of metals] through touch.[107]

[99] *bhṛgupātanam*. This seems to be a strengthened form of *bhṛgupatana* "a fall from a precipice."

[100] According to MW (citing Sanskrit lexicographers) these are milk, coagulated milk, butter, urine, and dung.

[101] = a *krośa*.

[102] = five *dhanus*.

[103] The Citrasūtras included *lākṣārasa* as an ingredient for the production of deep red (*atirakta*). See Nardi 2006: 127.

[104] *Koṭivedhin*. According to Monier-Williams, the compound refers to someone who can perform the most difficult deeds. I have translated it here in a more technical sense to refer to an adept who possesses the power of transmutation (*vedhana*). White has pointed out that "the Prabandhacintāmaṇi calls the [alchemical] elixir 'ten-million transmutation mercury' (*koṭavedhi-rasa*)," and the similar phrase seems to fit this context as well. See also RĀ 14.15–19, where *koṭivedhī* occurs in a sequential list of transmutations and Hellwig 2009: 449–450.

[105] *prākāraḥ* "rampart," "four-sided enclosed wall."

[106] This is probably a reference to the fourth-century CE Gupta ruler Candragupta I rather than the fourth-century BCE Mauryan emperor Candragupta. A Gupta-era city called "Chandra Gupta Patnam" was known to exist eighty miles south of Hyderabad, near Śrīśailam. Cf. Ellis 1876: 414–415.

[107] *sparśavedhakarā*. The translation follows RC (Hi. *vaha mudgavarṇ ki śilā sparś mātra se vedh karne vāli hai*).

137. A famous elephant is also there. It has been recognized that, having mounted it, a piece of straw is to be taken from its back. It shall become fully gold.

138. Also, he should dig into the earth, up to the length of the knees, in the elephant's vicinity to the north. [Stones][108] in the shape of plums (*jambūphala*)[109] are there. These cause the transmutation [of metals] when touched.

139. Or else he should dig, up to the length of the knees, below the elephant. He shall then obtain a triangular pill, supreme and perfected, capable of transmuting [metals] by the tens of thousands.

140. The Waters of the Gods (Devahrada), a bathing place for pilgrims,[110] is situated to the northwest of the Holy Mountain. There in the pool are mung bean-colored stones that cause the transmutation [of metals] when touched.

141. A lingam that resembles a turtle is there. It causes the best transmutations when touched. He should collect it after performing worship rites when [the moon is] in the Kṛttikā constellation, during [a lunar day of] Supūrṇā.

142. It is thought that fourteen kilometers[111] from there is a place named the Blue Forest (Nīlavana). There is a blue-colored lingam there that even looks like water.

143. Stones resembling horses, clouds,[112] and crows are there, which effect transmutation when touched. He should dig up to his knees inside of the god's sanctuary,[113] and carry them away.

144. A place named "Light of Guṇḍi" (Guṇḍiprabhā) is situated to the southwest of the Holy Mountain. Clay (*mṛd*), yellow in color, is to be taken from a hole in front of it. He should roast it inside a wrapper made of leaves.[114]

[108] "Stones" (*pāśānāḥ*) is not present in the original Sanskrit but seems to be implied based on similar passages in the text. RC also reads the verse in this way.

[109] On *jambū* as plums (and not the rose apple), see Wujastyk 2004.

[110] *tīrtha*.

[111] = two *gavyūti*s.

[112] *abhra*. This could also be a reference to *abhraka* "mica."

[113] *devagṛha*.

[114] *puṭair dahet*. This follows the reading in RC (Hi. *puṭ dekar jalaye*). "*puṭa*" can refer to an enclosure, which may be leaves, but also pots or a pit. It may also refer to the *process* of enclosed roasting, i.e., heating something that is enveloped or enclosed in something. The plural here may thus not refer to enclosure with multiple leaves, but to repeated enclosed roasting.

145. There is no doubt that glittering gold (*hema*) will come forth out of it. Bees shall also be there and give fruits just as before.

146. Likewise, in the expansive pilgrimage area named "Tambī's Town" [Tambīpura], he who is wise should always practice as before to attain the auspicious and famous fruit.

147. [A god] named "The Lord of Great Joy" (Mahānandeśvara) is to the southwest of Śrīśaila. He should go along the royal path in its proximity to the northeast.

148–149ab. A basin is in front of a god named the "Lord who is Black-Colored" (Kālavarṇeśvara). Mercury (*rasa*) that looks like a fire-fly[115] will be found inside it. It becomes gold (*kañcana*) by pouring the seven metals,[116] which have been heated, onto it.

149cd–150. The god Turaleśvara shall be to the northeast of the Holy Mountain. To his north is the celebrated and most excellent Hill of Umā. Splendid grasses stand at its summit in the shape of tridents.

151–152. Having dug beneath it as far as the hips, he should take away the blue soil. He should roast this using wood produced from the bahera myrobalan tree (*akṣaja*),[117] and should designate a portion for the god, should designate a portion for a guest, and a portion for Fire (Agni). A man who has eaten his own portion shall live for ten thousand aeons.

153. The famous Lord of Ten Millions (Koṭīśvara) is in the presence of the Lord of the Earth[118] (Ileśvara). In front of him mere flowers shall transmute [metals] when touched.

154–155. The supreme Lord of the Mountain (Acaleśa) is also there. He transmutes [metals] when touched. An auspicious gold

[115] *indragopaka*.

[116] Sanskrit alchemical texts differ from one another in their grouping of metals. A grouping of seven metals (*saptalohāni*) is rare, groups of six or eight kinds of metals being more common. The fourteenth-century *Rasādhyāya* (verses 160–161) lists a group of seven metals in the following order: silver, copper, iron, tin, lead, brass, and bronze (*rūpyaṃ tāmraṃ tathā lohaṃ vaṅgam nāgaṃ ca pittalam / kāṃsyaṃ ca saptalohāni*). On this passage and on other groupings of metals in medical and alchemical works, see Hellwig 2009: 248.

[117] This translation follows RC, which glosses *akṣa* as Hi. *baheṛā*, "bahera myrobalan." It may, however, also refer to "wood produced from rudraksha seeds" or to the tree itself (Elaeocarpus ganitrus).

[118] Or "Lord of Speech."

stone will be found a short distance of fourteen kilometers[119] to the south, called "The god who is the Immortal Lord" (Amareśvaradeva). He should place beef (*gomāṃsa*) or expensive meat[120] behind it.

156–157. After setting this [food] on fire using jujube wood (*badarīkāṣṭha*), the next morning it shall be seen to be like gold. After placing women's clothes, [stained] with menses and torn into pieces, atop this stone, he should place them in a bottle-gourd pot (*alābupātraka*). Metal becomes gold (*hāṭaka*) just by coming in contact with this.

158–159ab. At the Holy Mountain there are stones situated everywhere on the body of the earth. He should always consume twelve grams[121] of their powder together with the three fruits (*triphalā*).[122] These will slay old age after half a month, and he shall live as long as there are moon and stars.

159cd–160. The Lord Paṭāhakarṇa is to the southwest of the Kingdom of the Mountain, and a hundred and seventy meters[123] to his northeast is a stone that is elevated eighty-four centimeters[124] high. There he should dig eighty-four centimeters[125] deep.

161. He should fetch thirty of the stones that resemble the hollow of an ear, and, having made them as a hot as fire, he should sprinkle them with fluids produced from a pumpkin-gourd (*kuṣmāṇḍa*).

162. He who is intelligent should gather from inside it a portion of fresh butter for the god, one portion for a guest, and he should eat one portion.

163. The man will become faint and will arise as enlightened after a day and a night. The hero shall live for a myriad of years without grey hair or wrinkles.

[119] = one *yojana*.
[120] *mahāmāṃsa*, "great meat," "costly meat." MW notes this can also refer to human flesh.
[121] = one *karṣa*.
[122] Probably the three kinds of myrobalans.
[123] = a hundred *dhanus*.
[124] = two *hastas*.
[125] = two *hastas*.

164–166ab. The second step to this god transmutes upon touching. White and yellow stones are in the vicinity of this god. They all can be touched, there is no doubt. He should collect one. To the east of this god there shall be a well. In its midst are stones resembling frogs that transmute [metals] upon touching.

166cd–168ab. The famous Mountain of the Lingam (Liṅgaparvata) is to the north of this god, and the river Pūrvavāhinī will be to its north. A red-colored lingam is to the west of the river. There will be a pool with waters from the mountain that is capable of transmuting [metals] in an instant.

168cd–169. Caṇḍikā, called the famous "Goddess of the Piṇḍas" (Piṇḍādevī), is twenty-eight kilometers[126] to the south of the Great God.[127] A well stands at her northwest corner. Within it there are mung bean-colored stones.

170. These resemble yellow myrobalans (*pathyā*),[128] and they all transmute [metals] upon touching. The practice of their transmutation will be performed by the handling of them all.

171. He should crush the gross [stone] into fine pieces, and should prepare a crucible for it. The eight metals[129] should be stirred together within it.

172–173ab. All of it comes forth as gold (*svarṇa*). He should do this therefore according to his wishes. If it becomes thin[130] then he should place the melted eight metals down inside here and there. All of it will become gold.

173cd–176ab. The Pilgrims' Bath of the Lord of All (Sarveśvara) is situated to the northwest of the White-Jasmine God. While going in its southern direction along the water's path for seven kilometers,[131] a gourd (*ḍāṅgarika*) will be seen there. At the [water's] head black amla fruits (*dhātrīphala*) will be found.

[126] = two *yojanas*.
[127] Maheśa. This is possibly the same Maheśa as in verse 67.
[128] *mudgavarṇakāḥ tathā pathyānibha*. RC interprets *mudgavarṇaka* as "yellow-colored" (Hi. *pīt varṇ*) and *pathyānibha* as "green-colored" (Hi. *harā varṇ*).
[129] *aṣṭalohāni*: MW gives gold, silver, copper, tin, lead, brass, iron, and steel. See also Hellwig 2009: 248.
[130] *sūkṣma*. RC interprets this in a similar way, as in a fluid lacking in substance (Hi. *sūkṣm dravya*).
[131] = half a *yojana*.

The man should eat as much of them as he wishes, and in a week he shall be endowed with an adamantine body. Free from wrinkles and gray hair, he shall live for three hundred aeons.

176cd–178ab. To the south of there is the Great Forest[132] of Kākalerī. A plantain tree trunk is there, and there the adept should enter. He should go as far one and three-fourths a kilometer,[133] and there a pool of mercury (*rasa*) will be seen. After gathering the mercury in a pumpkin-gourd pot (*alābupātra*), it shall become capable of transmuting [metals] by the tens of millions.

178cd–179. A peacock that is filled with blue-colored stones will be there in the forest. A blue pool is in front of its beak, twenty-one centimeters[134] large. Dried bamboo (*vaṃśa*), after being cast inside of it, ought to become instantly like new.

180. Its leaves, fruit, and blossoms will come forth in a day—it will not be otherwise. After placing about eighty-one grams[135] of mercury (*pārada*) there [in the pool], he should take the same amount of water.

181. He should drink this with the same amount of honey. That very instant he shall become faint. After a moment he will be endowed with rulership and shall live for a thousand ages.

182. The Lord who Creates the Siddhas of the Pill (Guṭikāsiddhakeśvaraḥ) will be found to the southwest of the Mountain (Śaila). He should dig the ground in front of him up to about two meters[136] five forearms deep.

183. Stones in the shape of jujubes (*badara*) are there that bestow the power of flight. After collecting just one, he is to hold it in his mouth for traveling in the sky.

184. A rule will be said for all the yogas spoken of at the Holy Mountain: after first performing the bodily assignment

[132] JTĀ reads *mahānavam* while RC emends this to *mahāvanam* "the great forest," which I have followed here for intelligibility and due to the presence of *vane* in verse 178. GON is unclear but seems to present yet another possibility: *mahānanam*, "the great entrance."

[133] = half a *krośa*.

[134] = twelve *aṅgulas*.

[135] = one *prastha*. RC: Hi. *ser*.

[136] = five *hastas*.

(*nyāsa*) of a mantra, he should afterward recite the "Not Dreadful" (Aghora) [mantra][137] one hundred thousand times. By means of this he shall quickly receive this or that power.

185. He should invoke the Deity of the Field (*kṣetradevatā*). The deity, appearing before his eyes in a dream, there grants a boon.

The Rite of the Holy Mountain[138] is therefore to be carried out by steadfast devotees and adepts who have the supreme mantras, who have the worship of Śambhu as their highest aim, and who are daily guardians of the rites of worship, using the supreme recitations and the path of the highest among sacred transmissions. If this is not so, then their work of afflictions and rejected acts shall certainly be in vain.[139]

Bibliography

Âchârya, Jâdavjî Tricumjî, ed. 1913. Shri Nityanâthasiddha, *Rasayanakhanda: Fourth Part of Rasaratnâkara*. Bombay: "Nirnaya-sagar" Press.

Âchârya, Jâdavjî Tricumjî 1939. Nitya Nātha Siddha, Śrī *Rasâyanakhanda: Fourth Part of Rasaratnakara*. The Haridas Sanskrit Series 95. Banaras: Caukhambā saṃskṛta pustakālaya.

Ācāryaḥ, Yādavjī Trikamjī, ed. 1982. Śrīnityanāthasiddha. *Rasaratnākara-Rasāyanakhaṇḍaḥ Saṭippaṇa "rasacandrikā" Hindīvyākhyopetaḥ*. Edited by Vaidya Yādavjī Trikamjī Ācāryaḥ. Hindi commentary by Ḍā. Indradev Tripāṭhī. Haridās Saṅskṛta Granthamālā 15. Vārāṇasī: Caukhambā Amarabhāratī Prakāśan, 1982.

Ānandakanda. See Sastri 1952.

Balasubramaniam, R. 2009. "New Insights on Metrology during the Mauryan Period." *Current Science* 97 (5): 681–682.

Birch, Jason. 2018. "Premodern Yoga Traditions and Ayurveda: Preliminary Remarks on Shared Terminology, Theory, and Praxis." *History of Science in South Asia* 6: 1–83.

Bisschop, Peter Christiaan. 2006. *Early Śaivism and the Skandapurāṇa: Sects and Centres*. Groningen Oriental Studies 21. Groningen: Forsten.

Dagens, Bruno. 1984. *Entre Alampur et Śrīśailam: Recherches archéologiques en Andhra Pradesh*. 2 volumes. Pondicherry: Institut Français d'Indologie.

Ellis, R.R. 1876. "Sovereign." In *Notes and Queries: A Medium of Intercommunication for Literary Men, General Readers, Etc. 125*, edited by Doran, 414–415. Saturday, May 20, 1876.

[137] The Eighth Instruction does not itself specify what this mantra is, but a version is provided in the *Rasakhaṇḍa* of the *Rasaratnākara* (First Instruction, verses 32–33): *aghorebhyo'tha ghorebhyo ghoraghoratarebhyaḥ / sarvebhyaḥ sarvasarvebhyo namaste'stu rudrarūpebhyaḥ*. For another reference to a so-called Aghora mantra in the *Kubjikāmatatantra*, see Goudriaan and Schoterman 1988: 116–117.

[138] *śrīgirisādhanaṃ*.

[139] GON adds the colophon *iti śrīpāvatīputra śrīnityanāthasiddhaviracite rasaratnākare rasāyanakhaṃde śrīparvatasādhanaṃ nāma aṣthamopadeśaḥ / śrīrastu samāpto 'yaṃ rasāyanakhaṃḍaḥ*.

Fisher, Elaine M. 2019. "The Tangled Roots of Vīraśaivism: On the Vīramāheśvara Textual Culture of Srisailam." *History of Religions* 59 (1): 1–37.

Goudriaan, Teun, and J. A. Schoterman. 1988. *The Kubjikāmatatantra: Kulālikāmnāya Version*. Leiden: Brill.

Hellwig, Oliver. 2009. *Wörterbuch der mittelalterlichen indischen Alchemie*. Groningen: Barkhuis.

Mallinson, James. 2019. "Kālavañcana in the Konkan: How a Vajrayāna Haṭhayoga Tradition Cheated Buddhism's Death in India." *Religions* 10 (4): 1–33.

Mallinson, James. 2007. *The Khecarīvidyā of Ādinātha: A Critical Edition and Annotated Translation of an Early Text of Hathayoga*. Routledge Studies in Tantric Traditions Series. New York: Routledge.

Meulenbeld, G. Jan. 2000. *A History of Indian Medical Literature. Volume IIA*. Groningen: Egbert Forsten.

Monier-Williams, M. A. 1899. *Sanskrit-English dictionary: Etymologically and Philologically Arranged with Special Reference to Cognate Indo-European Languages*. Oxford: The Clarendon Press.

Nardi, Isabella. 2006. *The Theory of Citrasutras in Indian Painting : A Critical Re-Evaluation of Their Uses and Interpretations*. London: Routledge.

Olivelle, Patrick. 2013. *King, Governance, and Law in Ancient India: Kauṭilya's Arthaśāstra: A New Annotated Translation*. New York: Oxford University Press.

Rao, M. Rama. 1966. "The Tripurāṅtakeśvara Temple, Tripurāṅtakam." In *Select Kākatīya Temples*, 41–48. Tirupati: Sri Venkatesvara University.

Rao, M. Rama. 1969. *The Temples of Srisailam*. Hyderabad: Government Central Press.

Rasaratnākara. See Âchârya 1913; Âchârya 1939; Ācāryaḥ 1982.

Rasaratnākara. MS 5-3091 of the National Archives, Kathmandu, Nepal. Digitized by the Nepal-German Manuscript Preservation Project. Reel A 216/1.

Rasaratnākara (Caturtha: Rasāyanakhaṇḍa). R. No. 317 GON of Gujarat Ayurved University, Jamnagar.

Rasārṇava. See Tripāṭhī 1978.

Reddy, Prabhavati C. 2014. *Hindu Pilgrimage: Shifting Patterns of Worldview of Srisailam in South India*. New York: Routledge.

Roşu, Arion. 1969. "A La Recherche d'un Tirtha Énigmatique Du Dekkan Médiéval." *Bulletin de l'École Française d'Extrême-Orient* 55: 23–58.

Roşu, Arion. 1992. "Alchemy and Sacred Geography in the Medieval Deccan." *Journal of the European Āyurvedic Society* 2 (1): 151–157.

Sanderson, Alexis. 2012–2013, 2014. "The Śaiva Literature." *Journal of Indological Studies* 24–25: 1–113.

Sastri, S. V. Radhakrishna. 1952. *Anandakandam*. Tanjore: S. Gopalan. Digital input by Oliver Hellwig. http://www.sanskrit-linguistics.org/.

Sastry, P. V. Parabrahma. 1985. *Srisailam, Its History and Cult*. Guntur: Lakshmi Mallikarjunna Press.

Shaw, Richard. 1997. "Srisailam: Centre of the Siddhas." *South Asian Studies* 13 (1): 161–178.

Tripāṭhī Ayurvedācāryaḥ, Indradev, and Indradev Tārādattapantaḥ Āyurvedācāryaḥ, eds. 1978. *Rasārṇavam Nāma Rasatantram*. Vārāṇasī: Caukhambā saṅskṛta sīrīj āphis.

Vasudeva, Somadeva. 2012. "Powers and Identities: Yoga Powers and the Tantric Śaiva Traditions." In *Yoga Powers: Extraordinary Capacities Attained through Meditation and Concentration*, edited by Knut A. Jacobsen, 264–302. Leiden: Brill.

White, David Gordon. 1996. *The Alchemical Body: Siddha Traditions in Medieval India*. Chicago: University of Chicago Press.

Wujastyk, Dagmar. 2012. *Well-Mannered Medicine: Medical Ethics and Etiquette in Classical Ayurveda*. New York: Oxford University Press.

Wujastyk, Dagmar. 2017 "Acts of Improvement: On the Use of Tonics and Elixirs in Sanskrit Medical and Alchemical Literature." *History of Science in South Asia* 5 (2): 1–35.

Wujastyk, Dagmar, Suzanne Newcombe, and Christèle Barois. 2017. "Introduction." *History of Science in South Asia* 5 (2): i–xvii.

Wujastyk, Dominik. 1984. "An Alchemical Ghost: The Rasaratnâkara by Nâgârjuna." *Ambix* 31 (2): 70–83.

Wujastyk, Dominik. 2004. "Jambūdvīpa. Apples or Plums?" In *Studies in the History of he Exact Science in Honour of David Pingree*, edited by Charles Burnett et al. Leiden and Boston: Brill.

Wujastyk, Dominik. 2022. "The Surviving MSS of the Rasendramaṅgala of Nāgārjuna Siddha." Version v1.3, October 12, 2022. https://doi.org/10.5281/zenodo.7191411.

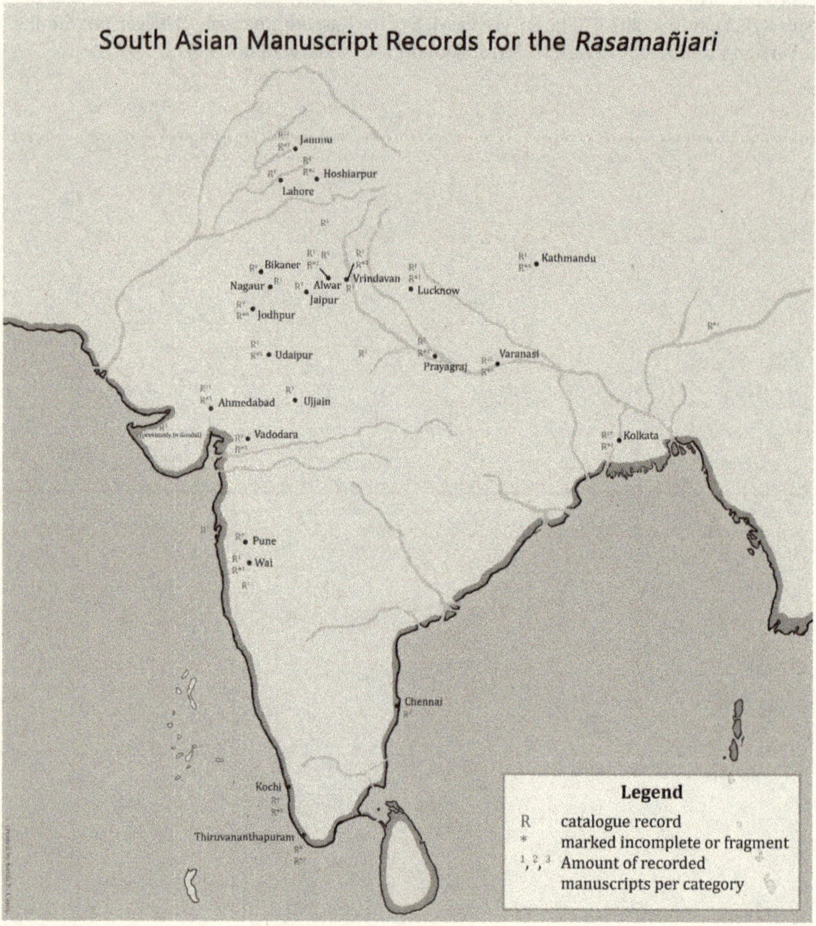

Map 6.1 South Asian manuscript records for the *Rasamañjarī*.

6
Potency, Virility, Sexual Pleasure
Chapters 9 and 10 of the *Rasamañjarī*

Patricia Sauthoff

The *Rasamañjarī* ("The Pearl of Mercury") by Śālinātha, a text in twelve chapters that dates to sometime in the early sixteenth century,[1] focuses on the medicinal usages of mercury and other alchemical substances. The text begins with a praise of mercury, the teacher and pupil, and the properties of mercury. It also examines the origins, usages, and purification processes of sulfur, diamonds, mica, gems, metals, plants, and other alchemical substances. Additionally, the text includes formulae for mercurials (*rasayoga*) and vitalization tonics (*rasāyana*), and recipes to cure eye disease, grey hair, and so on. Many of its *rasayoga* and *rasāyana* recipes are found only within the *Rasamañjarī*.

Chapter 9 of the *Rasamañjarī* focuses on human sexuality and reproductive medicines. The text offers recipes to delay ejaculation, increase potency, cure diseases, ensure impregnation, tighten the vagina, remove pubic hair, destroy sexual desire in others, prevent pregnancy, and avoid miscarriage. The topics of this chapter belong to the realm of "the production of virility" (*vājīkaraṇa*), the last of the eight subject areas of Ayurveda, focused on aphrodisiacs, virility, and improving progeny. As Alter demonstrates, virility in the context of *vājīkaraṇa* is synonymous with the ability to control semen.[2] This self-control includes both the containment and ejaculation of seminal fluid. Masculinity is measured both in the control one demonstrates through celibacy as well as through sexual potency. Further, the male must not necessarily choose only one to demonstrate his masculinity. Both the man with abundant children, especially sons, is just as manly as the celibate renunciate who practices ejaculation control as part of his yogic exercises.

[1] Meulenbeld 2000: 638.
[2] Alter 2008: 292.

Patricia Sauthoff, *Potency, Virility, Sexual Pleasure* In: *Indian Alchemy*. Edited by: Dagmar Wujastyk, Oxford University Press. © Oxford University Press 2025. DOI: 10.1093/9780197805831.003.0007

For the father, procreation ensures his continued line, whereas for the yogi death is a consequence of the fall of semen. Therefore, the preservation of semen allows the yogi to overcome death,[3] giving him longevity and resistance to disease and old age.[4] Semen here is not simply a sexual material but potency itself.

While the main concern of *vājīkaraṇa* is typically male sexual power, the majority of the *Rasamañjarī*'s ninth chapter focuses on the female body and its sexual health. Two sections, each thirty verses, are dedicated to recipes to improve female fertility and the appeasement of goddesses during pregnancy. Only the first twenty-four of chapter 9's seventy-eight verses focus on the male body. Though largely concentrated on medicines for the female body, these medicines often focus on sexualizing the body for male pleasure through the tightening of the vagina (37), lifting of the breasts (44–45), the removal of pubic hair (38–39), and mixtures to subjugate women in order to make them "like a slave" during sex (31–36). Menstrual blood is first mentioned with ashes and a corpse (40–41), bringing to mind the tantric imagery of fierce deities and transgressive ritual behaviors.[5] Only two verses focus on abnormally heavy menstruation (25–26), and one offers relief from uterine pain (49). These appear in a section that also includes recipes to ease spleen diseases (27), urinary stones (28), and polyuria (29). The text prescribes various concoctions to ensure pregnancy and, especially, to ensure the birth of a son (50–78). Additionally, several verses offer contraceptive recipes (46–48). Despite the contraceptive recipes, there is no mention of abortion in the *Rasamañjarī*. This is unsurprising as abortion recipes do not begin to appear in medical texts until the late sixteenth or early seventeenth century. The *Carakasaṃhitā* and *Suśrutasaṃhitā* both mention abortion but do not provide information on how to perform one.[6] The late sixteenth- or early seventeenth-century *Yogacintāmaṇi* and the late seventeenth-century *Vaidyavallabha* are the first medical works to provide recipes using abortifacients such as long pepper, false black pepper, and hibiscus.[7] While the *Rasamañjarī* ostensibly only provides recipes to prevent conception, the ingredients in these recipes are known abortifacients. Much

[3] Mallinson 2018: 188.
[4] Dash and Basu 1968: 10.
[5] For a discussion of Meulenbeld's claim that the chapter describes the tantric magical acts (*ṣaṭkarman*), see Sauthoff 2023.
[6] Wujastyk 2012: 144.
[7] Ibid.

more focus is on impregnation and the prevention of miscarriage. The rites of chapter 10 center on protecting the newborn child from *yoginīs*, fierce female demons who cause illness and suffering, especially in children.

In the *Rasamañjarī*, the female experience of sex is completely absent, though there is an acknowledgment of female sexual pleasure. Verses 19–20 teach that a man should make a concoction that will bring about multiple orgasms in his female partner(s). Male sexual pleasure too is rarely mentioned in the text, with the main focus being on the retention or emission of ejaculate and reversing impotence or causing it in another. This demonstrates that while masculinity can be defined by the man who can control his ejaculate and/or father children, femininity is defined by physical youth, beauty, and the ability to bear children. Elsewhere in Sanskrit literature we do find discussions of female pleasure and orgasm. The fifteenth- or sixteenth-century *Anaṅga Raṅga*, which draws on the *Kama Sūtra* and includes recipes for mercurial aphrodisiacs, discusses these issues at length. However, they are the purview of the male, whose function in non-procreative intercourse is to bring both himself and his partner to satisfaction. The feminine is defined in opposition to the masculine. Further, masculinity can be reaffirmed by bringing about orgasm and pleasure in his partner.

Femininity is, however, not simply about procreation, despite the *Rasamañjarī*'s focus on reproduction. What the male gaze of ancient and medieval texts tell us about women is that they are objects of sexual desire for men, that reproduction is key, but also that women are dangerous. One must both protect women and appease feminine wrath. This is most clear in descriptions of the *yoginīs*. In early tantric literature[8] these flying, flesh-eating female creatures wreak havoc if not appeased through the offerings of blood and flesh. The *Rasamañjarī* does not offer specific recipes for their appeasement but instead warns about the symptoms of illness brought by the *yoginīs* to the unborn child. Women too are at risk from the *yoginī*, gaining no special protection through their femininity. The *Rasamañjarī* opens its section on pediatrics citing Rāvaṇa's *Bālatantra*. The text and *yoginīs* found in the *Rasamañjarī* differ from those found in several published editions of the *Bālatantra*. However, most of these *yoginīs* appear among the sixty-four found in the pre-eleventh-century *Kriyākālaguṇottara*.[9] Explanations for illness, especially in children, have long had a place in both tantric and

[8] Törszök 1999: xv.
[9] Slouber 2016: 9.

ayurvedic works. The *Suśrutasaṃhitā*'s (SS) Uttaratantra 27 lists nine malignant forces[10] that affect infants whose mothers or nurses did not follow the instructions for childcare and who failed to perform the proper benedictory rites laid out in the Śārirasthāna (chapter 10).[11] Among the demons (of the class seizers or *grahas*) listed in the SS both Pūtanā and Śakunī also appear in the *Rasamañjarī.* White also notes two instances of goddesses who attack children, one tantric and one ayurvedic. The first is from the ninth-century *Netratantra*, which retells a myth from the possibly seventh-century ayurvedic pediatric manual *Kaśyapa Saṃhita*,[12] though he notes the latter appears to be unknown to the commentator of the former.[13] In the *Kaśyapa Saṃhita* we find three of the sixteen *graha*s mentioned in the *Rasamañjarī*: Śakuni, Pūtanā, and Bhadrakālī.[14]

While the vocabulary of the *Bālatantra* quotations found in the *Rasamañjarī* differ from those of the published edition, the grammatical structure remains intact. The refrain "*divase māse, varṣe*" appears before the names of the various *yoginIs*, telling us the day, month, or year in which she possesses the child. The *Bālatantra* also describes its *yoginī*s as the sixteen sisters of Rāvaṇa, who were granted a boon by Śiva, who allows them to eat newborns' flesh and attributes childhood diseases to this possession.[15]

The *Rasamañjarī* is not unique among alchemical texts for its focus on issues of sexuality and reproduction. Recipes for aphrodisiacs are found in the *Rasendramaṅgala* (chapter 4), *Rasaprakāśasudhākara* (chapter 12), and the *Rasaratnākara*'s *Rasāyanakhaṇḍa* (chapter 6). Similarly, the *Rasendramaṅgala* (chapter 4), *Rasaprakāśasudhākara* (chapter 13), and the *Rasaratnākara*'s *Rasāyanakhaṇḍha* (chapter 7) discuss the suppression of ejaculation (often, but not always, called *vīryastambha*). Likewise, alchemical formulations for sexuality do not only appear in *rasaśāstra* texts. For example, chapter 6 of the *Anaṅga Raṅga* has mercurial recipes for bringing women under the sexual spell of a man, the delay of male ejaculation, and the hastening of female orgasm.

[10] Bhishagratna 1916: 141.
[11] Bhishagratna 1911: 225.
[12] See Meulenbeld 2000: 41 for dating issues.
[13] White 2003: 146–147.
[14] Meulenbeld 2000: 32–35.
[15] White 2003: 51.

Notes on the Text and Translation

In consulting different published editions it becomes clear that there are variations within editions. The 1995 edition published by Caukhambhā Oriyaṇṭāliyā is an exact copy of the 1907 edition. The Digital Corpus of Sanskrit relies on a 2003 printing also by Caukhambhā Oriyaṇṭāliyā but is different in verses 21–32 from the 1907/1995 editions. The only version of the 1927 edition I have been able to consult is missing verses 17–26. However, verses 27 onward conform to 36 onward in the 1907 edition. Aside from these minor variations, the published editions of the *Rasamañjarī* are largely consistent in verse order and wording. I have relied mostly on the 1907 edition, the only version to include the *Bālatantra* section as a separate chapter. Unfortunately, time and resources did not allow for the study of *Rasamañjarī* manuscripts, of which nearly 200 are spread across the subcontinent, with nineteen alone split between Kochi and Thiruvananthapuram, thirty-two in Varanasi, and twenty-four across Jammu, Hoshiarpur, and Lahore. Clearly, this was a widely disseminated text, which may reveal variations, especially within the *grahas*.[16]

The most difficult part of this translation has been the word choice for anatomy, sexual attitudes, and medical conditions. Consultation of Victorian-era dictionaries leaves us with out-of-date terminology such as "wanton," while the more current and morally neutral "sexually active" loses the pejorative sense that does exist in the original Sanskrit. The translation must be honest to the original while remaining modern and readable. It must also acknowledge the male perspective of both the author of the text and the alchemist who prepares the remedies. This task is easier with out-of-fashion terms such as "slut" (*raṇḍā*) but more difficult with the antiquated "barren" (*vandhyā*). I have avoided terms such as "prostitute," "sex worker," and "courtesan" as a translation for *veśyā* as each term reduces the complexity of women's sexuality to a term that maps the present onto the past. Instead I have used "sexually active" for *raṇḍā* and "unmarried and promiscuous" for *veśyā* to hint at the complex nature of female sexuality and prostitution in medieval India.[17]

[16] For example, Lipsitā is sometimes called Nandikā, Acintā called Acintyā, and Śavarī called Sumukhī. Meulenbeld 2000: 638.

[17] See Mitra 2020: 46–57 for a discussion of philology and prostitution in medieval India and colonial-era translation.

The *Rasamañjarī* does not define binary experiences of fertility with the use of a prefix of negation. Therefore, I have chosen more neutral terminology, despite the current medical normativity of the word "infertility." Instead, I have chosen to use the terms I would use with other women in a non-medical setting. Rather than "infertile," I use "a woman who has difficulty conceiving." This is still imperfect but I hope conveys a sense of optimism in the medical procedures that have inspired women to undertake treatments throughout history.

Within Sanskrit literature, the womb (a masculine noun in Sanskrit) is metaphorically often connected with the earth (typically a feminine noun in Sanskrit). Penis (*liṅga*) is straightforward while vulva (*yoṇimukha* or "vaginal mouth," "the orifice of the womb") utilizes more metaphorical naming, reminiscent of the American English colloquial term "lips" for the labia. One may be inclined to read patriarchal undertones to the masculine declension of womb (*garbha*) and *yoni*, a term that acts as a catch-all for the female anatomy, translated variously as womb, uterus, vulva, vagina, and female organs of generation. Penis (*liṅga*) is occasionally feminine but most often a neuter noun. However, the inconsistency demonstrates that the gender of anatomical terms is irrelevant. Most Sanskrit terms for anatomy have myriad other meanings depending on context and the declension remains the same regardless of meaning.

Rasamañjarī, Chapter 9

1. Having mixed camphor, borax, quicksilver, the juice of the hummingbird tree, and honey equally, he should coat the penis, remaining thus for three hours.
2. Then, he should rinse the penis. He will delight a suitable and lovely woman. Nagarjuna correctly proclaimed this to inhibit men's ejaculation (*vīryastambhakara*).
3. A lizard's tail tip ring is worn over the little finger, wrapped with fixed threads: a man will not ejaculate.
4. Having rubbed lotus seeds with honey, he should rub it on the navel. So long as it remains smeared there, he does not ejaculate.
5. He should take a sparrow egg and mix it with fresh butter. He should smear both feet with this. This suppresses ejaculation. So long as he does not touch the earth, he will not ejaculate.
6. He should collect the tip of the right tooth of a forest boar. Whenever he has bound it above at the hip, suppression of ejaculation occurs.
7. He should get a black colored snake, called "lizard."[18]
8. He should retain the bone of this on the hip. Through loosening it he ejaculates. This is declared in the *Siddhayoga*.[19]
9. Chaff flower root is consecrated by the Monday mantra (*somavārābhimantrita*). At dawn on Tuesday, he spreads and binds it to the hip, he does not ejaculate.
10. Forty-eight grams[20] of poppy[21] and a decoction of dry ginger mixed with sixteen parts of treacle is drunk at night. He does not ejaculate during sex without [eating or drinking] something sour.
11. He should eat elephant foot yam and tulsi root with betel. A man does not emit a single sperm, there is no doubt.
12. He should smear the penis with boar lard and with honey. The man's penis will become thick, solid, and long.

[18] Possibly a non-venomous snake according to Ullrey 2016: 269.

[19] Not found in the *vīryastambha* section of the *Siddhayoga* of Vṛṇḍa.

[20] = four *karṣa*.

[21] The term used here, *khasa*, is not found in any dictionary as poppy. However, *khaskhasa* and *khasatila* are poppy.

13. And he should mix the filament of a white lotus with and equal amount of honey. He should cover the penis with it, at a time other than during sex. There is certainty.
14. When the moon is in Anurādha[22] the heart fern roots are permanently buried at the place of sex. This produces the destruction of manhood [impotence].
15. Adding the powder of six poison nuts steeped in goat urine to food and drink at night causes impotence among men.
16. Impotence is destroyed by eating sesame seeds and powder of small caltrops cooked with goat's milk and mixed with cold honey.
17. A man who has a circle of hair around the navel should smear [it] with honey. Because of the smearing the navel with this he is freed from the bond of impotence.
18. The juice of pressed jasmine is a great aphrodisiac (*mahādrāvin*) on its own, and even better when it is joined with honey, camphor, and mercury.

19–20. Now, having crushed equal measures of fresh ginger, sulfur, Indian mustard, and also borax, he should add it in equal amounts into lime juice and water. He should leave this for seventy-two minutes. He should hold it in the hand. When women smell it, they orgasm as many as five times.

21–22. A paste of iron pyrite, mercury, shilajit, chebulic myrobalan, false black pepper, thorn apple, cannabis, and nutmeg leaf should be soaked with the distillates of ashwagandha and small caltrops one by one. One should take a seventh of each of these with honey and ghee.

23–24. Having eaten wheat flour milk pudding prepared in the milk of cows with grown calves, mixed with white sugar, honey, and clarified butter, even one who is visibly aged has sex with ten wives. Celibacy is to be observed for up to twenty-one days.

25. Ninety-six grams[23] of mica are mixed with white sugar and milk. A woman who eats ghee and milk is freed from abnormally heavy menstrual bleeding.

[22] The seventeenth of the twenty-eight lunar mansions (*nakṣatra*). It is unclear why Anurādhā is mentioned here as instructions relating to the *nakṣatras* are not found elsewhere in the text.

[23] = two *pala*.

26. One should properly crush ashoka tree bark distillate, ghee, and cold milk in the morning to calm the abnormally heavy menstrual bleeding of women.
27. Pearl and long pepper distillates with milk or with salt remove spleen disease.
28. White gourd melon distillate, combined with barley alkali and asafoetida, drives away urinary stones.
29. The distillate of sickle senna root ground with rice-water, drunk at the coming of daybreak, carries away polyuria.
30. Ashwagandha distillate prepared together with milk, drunk by a woman in her fertile period, causes conception.
31. Having smeared a combination of honey and rock-salt mixed with pigeon feces on the penis, it makes a woman like a slave during sex.
32. Having smeared monkey semen, camphor, quicksilver, saffron, gold, and honey on the penis, it makes a woman like a slave during sex.[24]

33–35ab. One should mix equal parts of sandalwood, valerian, costus, perfume cherry, Indian rose chestnut, long pepper, and tamarind. He should give [portions] of that preparation together with the five excretions for seven days. It should be given to either woman or man in food and drink. As long as she lives, she will always live as his slave in soul and in wealth.

O, salutations to you, Umāmaheśvara. O, greetings Mohinī. In great fortune, Śiva! By this, having done the aforementioned seven times, one should be joined. One is subjugated.

35bc–36. One should mix equal parts shankapushpi, croton, fennel leaf, and orchid tree mixed with *giri*. One should give this in food and drink with the five excretions for five days. It is the ultimate [potion for] subjugation.

[24] Anon 1927 and Miśra 2003 have a variant reading for verses 21, 24, 25 and conform to Mukundarām1907/Miśra 1995 from 33 onward.

Joining pulverized honey with the leaves and fruit of the chinaberry tree and smearing it on the penis, he should quickly embark on sex with women. 21

The name is situated in the middle of the circle. On the outside *hrīṃ* four times. Outside of this, *klīṃ* fourfold. Having written this with red arsenic, it is complete. 24

In union with the beautiful Nakṣatra, situated in honey, certainly he should quickly and forcibly attract a woman, there is no doubt. 25

37. Having first said the aforementioned mantra seven times, one should administer it, having bathed the genitals with emblic myrobalan bark. Even an old woman becomes impassioned and makes love like a young woman.
38. Through the daily application of powdered orpiment on the hair line extending from the pudenda to the navel together with water, all the hair falls out. One performs this marvelous miracle.
39. One should burn flame of the forest, tamarind, sesame, mung bean, and shankapushpi. Chaff-flower with peepal or red arsenic. It quickly makes the [pubis] hairless because of the anointing with orpiment powder.

40–41. One should take the dust stirred up by a donkey and the ash from the body of a corpse together with menstrual blood. Having mixed this together one should throw this onto a bed at night or on a seat. It is stated in the *Mālatīmata* (The View of Jasmine) that aversion will be produced immediately.

42–43. I will speak of another excellent potion causing extreme aversion: collect the dust of two opposing dogs fighting with one another, it will kill the sexual pleasure of a woman. Truly, aversion will arise, in this there is no doubt.

44. After lotus seed is eaten with white lotus water, a woman's pair of breasts become very firm in a month.
45. Sesame oil boiled with globe thistle powder decoction perks up the breasts of one whose youth has passed.
46. Having taken crushed amaranth roots with rice water for three days at the end of the fertile period, a woman becomes unable to conceive.
47. Crushed China rose flowers with sour gruel, when taken by a woman for three days during the fertile period, will make the woman unable to conceive after being absorbed.
48. Taking white thorn apple and jasmine flower and placing them together at the hip prevents conception in women who are sexually active or unmarried and promiscuous, and so on.
49. If the opening of the vagina is incensed with neem wood, a woman who has sex at the end of her fertile period is free from uterine pain.

50. Through partaking of Indian rose chestnut together with cow's ghee and eating milk rice, she becomes pregnant with a son during her fertile period.

51. Citron seeds boiled in milk together with ghee make even a woman who has difficulty conceiving pregnant by drinking this.

52. One should grind citron seeds with aloe.[25] This should be given with milk. She certainly becomes pregnant.

53. Long pepper, ginger, black pepper, and saffron should be drunk together with ghee. This is the best means of providing fertility.

54. Having drunk extracted flower blossom of citron (*mātuliṅga*) and crushed aloe root, together with ghee and milk at the end of the fertile period, she becomes pregnant.

55. For women whose menstrual flow has not arrived during the fertile period, this treatment is to be done again and the period should arrive.

56–57. The physician should make a pill of long pepper, alkali, false black pepper, wood apple fruit, Tellicherry bark, and bottle gourd seed. Indian globe thistle mixed with milk on the cervix of any suitable woman brings menstruation in or three, five, or seven nights, not otherwise.

58–60. I will describe another excellent compound through which a woman should become pregnant. One should make [a medicine], using equal parts of licorice grass, lodhra, and also Tellicherry bark seed, ghee, shala tree resin, honey,[26] fig tree, and roots of moringa tree. Then, having crushed the ingredients, he should cook them with goat's milk. A woman should drink this for as many as seven nights until she menstruates.

61–62ab. Afterward, one should put this great medicine in her cleansed vagina. Mixed with aloe, one should administer it through the nose or as a drink. With this she obtains a virtuous son who is honored by the gods.

[25] Here *kumārī* can also refer to cardamom, aloe, butterfly pea, or jasmine.
[26] = *mākṣika*, which may also mean pyrites.

62cd–63ab. One should administer milk joined with garlic in the nose or as a drink. One should also give ashwagandha powder with goats milk.

63cd–64. Having reduced an equal share of barley alkali, false black pepper, guduchi and pepper[27] to a good powder and having drunk this, she should obtain a son, there is no doubt.

65–67. One should mix twelve grams[28] each of asafoetida, durva grass, pomegranate, rock-salt, three pungent spices, dill, cobra saffron, asparagus, licorice, and jasmine. These should be given together with milk to both the woman and the man. For three days one should eat rice, threshed grain, and milk. Once eaten, women should become pregnant, there is no doubt.

68–71ab. Milkweed root, perfume cherry, safflower, cobra saffron, country and Indian mallow, and goat's milk are drunk for three days. They purify the female reproductive system. Then, one should administer the great medicine. One should give ground blue lotus, Indian valerian, costus, licorice with sandal, goat's milk for five days. One should give her a beverage of mandrake mixed with cow's milk. With this, she obtains a son marked by wealth and very wise.

71cd–72ab. When her menstrual blood is gone and she is happy and pure, a woman should be given mandrake with milk as a nasal treatment or as a drink, and she will obtain a offspring.

72cd–73. Crown flower, white-flowered eggplant, oleander, white pea flower, mandrake, and poison nut, mixed with cow's milk, are to be given as a drink or given through the nose. She should obtain pregnancy from pleasurable sex.

74–75. One stops miscarriage by consuming potters clay with honey and goat's milk or else with white pea flower pericarp. Sal tree, candied sugar, velvet leaf, and elephant foot yam, mixed with honey and eaten instantly prevents miscarriage.

76. Pregnancy is well protected by drinking equal parts refined sugar and rice grain powder mixed into a decoction of cluster fig root.

[27] Piperaceae aurantiacum.
[28] = one *karṣa*.

77. A woman should easily obtain offspring by drinking citron roots, licorice mixed with honey together with ghee.
78. One should smear elephant foot yam crushed with sour gruel and the flame lily on the feet. The pregnant woman quickly gives birth.

Chapter 10[29]

1. I will succinctly explain pediatrics, as told by Rāvaṇa. This [includes] ritual object, directions [to face in offerings], mantra, and the characteristics of possession.
2. On the first day, month, or year, the yoginī named Nandā possesses a child. He has the following symptoms: [missing].

3–4. On the second day, month, or year, the yoginī named Sunandā possesses a child. First fever arises. The symptoms of one possessed by her are contractions of the feet and hands, eye disease, vomiting, being fearful, and being emaciated.

5–6. On the third day, month, or year, the yoginī named Pūtanā possesses a child. [Possession by her is] characterized by breaking limbs, fever, loss of appetite, crying, swelling of the head or neck, vomiting, and so on. One should cast oblations in the afternoon and in the west for five nights.

7–8ab. On the fourth day, month, or year, the yoginī named Biḍālī possesses a child. Possession by her [is characterized by] eye pain, fever, and loss of appetite. When pale and [subjected to] that which is called "breaking the limb" (*gātrasphoṭana*), and so on, one should cast offerings.

8cd–9ab. On the fifth day, month, or year, the yoginī named Pūtanāmātṛkā possesses a child. [Possession by her is] characterized by [missing].

9cd–11ab. Now, on the sixth day, month, or year, the yoginī named Śakunī possesses a child. Possession by her is characterized by cough and panting, loss of appetite, fever, contractions of the hands and feet, and so on, and eye pain. One should cast an oblation to her for five nights in the western direction.

29 I would like to thank Michael Slouber for his notes on my translation of this chapter.

11cd–12. On the seventh day, month, or year Śuṣkāśivā possesses a child, characterized by crying, trembling, fear, dryness, and so on. When five tributes have been given in the western direction the child is happy.

13–14. On the eighth day, month, or year Jṛmbhakā possesses. First, fever arises, and head ache, eye diseases, and exertion of optic nerve because of her possession. One should give an offering to her affixed in the southern direction.

15–16. On the ninth day, month, or year, the yoginī named Acintā possesses a child. Possession by her is characterized by limb contraction, head and eye pain, vomiting, crying, and so forth. One should give an offering to her, practicing in the northern direction.

17–18. On the tenth day, month, or year the yoginī known by the name Kāpālikā possesses a child. Possession by her is characterized by crying, trembling, vomiting, fever, weakness, sickness, and eye pain. One should dispatch an offering for five nights, seeking refuge in the Eastern direction.

19. On the eleventh day, month, or year, Lipsitā seizes. Possession by her is characterized by limb tremors and strong fever.

20–21. On the next day, month, or year, the yoginī named Pītalī possesses a child, who cries, agonizes, has a fever, who sleeps and who has lost its voice, who is jaundiced, and who is voided of food through vomiting. One should give an offering for seven nights in the northern direction.

22–23. On the next on the day, month, or year the one named Bhadrakālī possesses a child, who has a fever, trembling of the left hand, agony, loss of appetite, melancholy, yellow body, and struggling. One should give an offering to her, seeking refuge in the Eastern direction.

24–25ab. On the day, month, or year of Indra, the yoginī named Tārā possesses a child with fever, desiccation, a strong lack of appetite, eye pain, and voice change. One should bring an offering to her in the west.

25cd–26. On the [sixteenth] day, month, or year, the yoginī named Śarvarī possesses a child. Possession by her is characterized by panting, cough, loss of appetite, fever, and so forth. One should dispatch an offering in the south.

27–28ab. On the [seventeenth] day, month, or year the yoginī Kumārī seizes, acting through trembling eyes, fever, desiccation, and so on. One should dispatch an offering for seven nights, seeking asylum in the southwestern direction.

28cd–29. The knowers of mantras should bathe the child in the west with ritual water. The devī's own form on the earth is drawn to both river banks. Having done and accomplished the rite, [a woman receives] a prince or unbroken menstrual flow, and so on.

30–31. First, serving pills of laḍḍu flour and a ball of coagulated milk, [marking] the signs of the four castes, and burning sandal flowers, one should worship all diseases with all one's power in the afternoon. The offering should be given by always disclosing the name [of the yoginī].

32. Oṃ, greetings to the blessed goddess [name of goddess], release the child, accept the offering. Svāhā. This is the mantra for ritual offerings at all times.

Bibliography

Alter, Joseph. 2008. "Ayurveda and Sexuality: Sex Therapy and the 'Paradox of Virility'." In *Modern and Global Ayurveda: Pluralism and Paradigms*, edited by Dagmar Wujastyk and Frederick M. Smith. Albany: State University of New York Press.

Ananga Ranga. See Ray 1964.

Anon. 1927. *Rasamañjarī: vaidyak Sālināthaviracita: Mukhiyājī Raghunāthajī Srīkṛṣṇalāla Mathurānivāsīkṛtabhāṣāṭīkāsahita*. Kalyāna: Gaṅgāviṣṇu Srīkṛṣṇadāsajī.

Bhishagratna, Kaviraj Kunja Lal. 1911. *An English Translation of The Sushruta Samhita With a Full and Comprehensive Introduction, Additional Texts, Different Readings, Notes, Comparative Views, Index, Glossary and Plates in Three Volumes.* Edited by Kaviraj Kunja Lal Bhishagratna. Volume 2. Nidāna-Sthāna, Śārīrasthāna, Chikitsita-Sthāna and Kalapa-sthāna. Calcutta: Published by the Author.

Bhishagratna, Kaviraj Kunja Lal. 1916. *An English Translation of The Sushruta Samhita With a Full and Comprehensive Introduction, Additional Texts, Different Readings, Notes, Comparative Views, Index, Glossary and Plates (in Three Volumes). Edited by Kaviraj Kunja Lala Bhishagratna.* Volume 3. Uttara-Tantra. Calcutta: S.L. Bhaduri.

Carakasaṃhitā. See Sharma, Priya Vrat 1994.

Dash, Bhagwan, and R. N. Basu. 1968. "Methods for Sterilization and Contraception in Ancient and Medieval India." *Indian Journal of History of Science* 3 (1): 9–24.

Mallinson, James. 2018. "Yoga and Sex: What is the Purpose of *Vajrolīmudrā*?" In *Yoga in Transformation: Historical and Contemporary Perspectives*, edited by Karl Baier, Philipp André Maas, and Karin Preisendanz, 181–222. Göttingen: V&R unipress.

Meulenbeld, Gerrit Jan. 2000. *History of Indian Medical Literature*, Volume IIA. Leiden: Brill.

Miśra, Siddhinanda, ed. 1995. *Rasamañjarī* of Śālinātha. *Siddhipradā hindīvyākhyāsahita.* Varanasi: Caukhambhā Oriyaṇṭāliyā

Miśra, Siddhinanda, ed. 2003. *Rasamañjarī* of Śālinātha. *Rasamañjarī: Siddhipradā hindīvyākhyāsahita*. Varanasi: Caukhambhā Oriyaṇṭāliyā.

Mitra, Durba. 2020. *Indian Sex Life: Sexuality and the Colonial Origins of Modern Social Thought*. Princeton: Princeton University Press.

Mukundarām, Nārāyaṇaprasāda, ed. 1907. *Śrīvaidyanāthatanayaśālināthanirmita Rasamaṃjarī. Miśrakulāgragaṇyapaṇḍitanārāyaṇaprasādamukundarāma bāṃsarelī tathā lekhimapuranivāsikṛtabhāṣāṭīkāsahita*. Mumbaī: Gujarātī Priṇṭiṅga Presa.

Pādhye, H. K., ed. 1894. *Siddhayoga of Vṛṇḍa. Śrīmad-Vṛnda-praṇīto Vṛndamādhavāparanāmā Siddhayogaḥ. Śrīkaṇṭhadattaviracitavyākhyākusumāvalyākhyaṭīkāsametaḥ*. Pune: Ānandāśramasa Press.

Rasamañjarī. See Miśra 2003; Miśra 1995; Mukundarām 1907; Anon 1927.

Ray, Tribidnath. 1964. *Kalyanamalla. Ananga Ranga. Translated and edited by Tridibnath Ray*. New York: Citadel Press.

Sauthoff, Patricia. 2023. "Tantric Dimensions of Alchemy". In *The Oxford Handbook of Tantric Studies*, edited by Richard K. Payne and Glen A. Hayes, 301–320. Oxford: Oxford University Press.

Selby, Martha Ann. 2005. "Narratives of Conception, Gestation, and Labour in Sanskrit Āyurvedic Texts. *Asian Medicine* 1 (2): 254–275.

Sharma, Priya Vrat, ed. 1994. *Caraka Saṃhitā. Text with English Translation*. Varanasi: Chaukhambha Orientalia.

Siddhayoga. See Pādhye 1894.

Slouber, Michael. 2016. *Early Tantric Medicine: Snakebite, Mantras, and Healing in the Garuda Tantras*. Oxford: Oxford University Press.

Suśrutasaṃhita. See Bhishagratna 1911 and 1916.

Törszök, Judit. 1999. "The Doctrine of Magic Female Spirits: A Critical Edition of Selected Chapters of the *Siddhayogeśvarīmata(tantra)* with Annotated Translation and Analysis." PhD diss., Oxford University.

Ullrey, A. M. 2016. "Grim Grimoires: Pragmatic Ritual in the Magic Tantras." PhD diss., University of California–Santa Barbara. Ullrey_ucsb_0035D_13314.pdf.

White, David Gordon. 2003. *Kiss of the Yogini: "Tantric Sex" in Its South Asian Contexts*. Chicago: University of Chicago Press.

Wujastyk, Dagmar. 2012. *Well-Mannered Medicine: Medical Ethics and Etiquette in Classical Ayurveda*. Oxford: Oxford University Press.

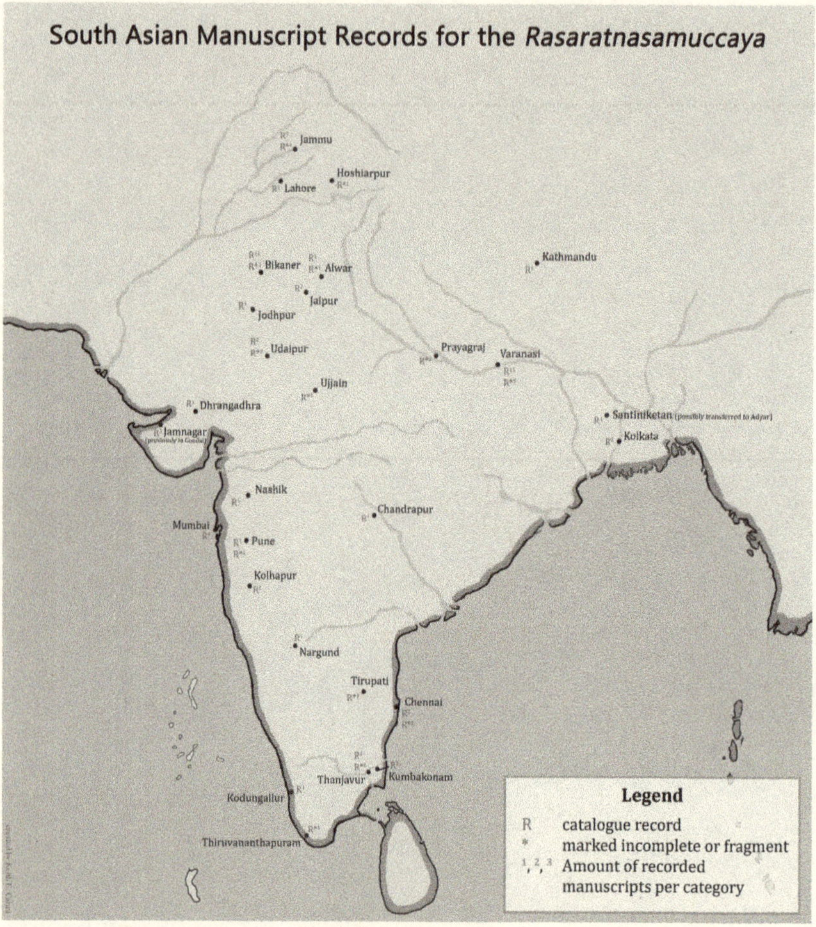

Map 7.1 South Asian manuscript records for the *Rasaratnasamuccaya*.

7
Medicine and Alchemy
Chapter 19 of the *Rasaratnasamuccaya*

Priyanka Chorge and Dagmar Wujastyk

In this chapter, we will examine the role medicine plays within the Indian alchemical tradition. A chapter from the fifteenth-/sixteenth-century *Rasaratnasamuccaya* (Collection of Mercury Jewels), one of the most widely read Sanskrit alchemical works, will serve to illuminate the approach to medicine that developed in alchemical milieus.

In his pioneering work *History of Hindu Chemistry* (1902), P. C. Ray introduced the idea that Indian works on alchemy divide into ones belonging to an earlier tantric period, and ones belonging to a later iatrochemical one. He suggested that

> During the Tantric period, with its system of the "Philosophy of Mercury," a vast mass of chemical information was accumulated, which was pressed into signal service in the period immediately succeeding it—the Iatro-chemical Period of India. The prominent feature of the former lies in the search after the elixir vitae and the powder of projection . . .; whereas in the latter these fantastic and extravagant ideas, impossible of realisation, had subsided into something more practical and tangible. The numerous preparations of mercury, iron, copper and other metals, although they could not secure immortality or revive the dead were found to be helpful accessories in medicine. (Ray 1902: lii)

This idea of a division between earlier works characterized by their tantric character, and later ones defined by their emphasis on medicine, has been echoed but also refined in later studies on Indian alchemy.[1] The exclusive identification

[1] See, for example, White 2013: 490, who traces this development from earlier to later alchemical works. White uses a different terminology, differentiating between an earlier tantric elixir alchemy

Priyanka Chorge, *Medicine and Alchemy* In: *Indian Alchemy*. Edited by: Dagmar Wujastyk, Oxford University Press.
 DOI: 10.1093/9780197805831.003.0008

of the search for an elixir vitae with a tantric milieu is challenged by the lack of tantric features in the tenth-century *Rasahṛdayatantra*: while it certainly describes the aim of achieving immortality in a living body and the attainment of a Śiva-like state, it does not feature tantric elements such as the use of mantras or diagrams (*yantra*), and it does not refer to deities specifically associated with tantric practice or to tantric lineages, as the twelfth-century *Rasārṇava* does.

The question of the presence of medicine in alchemical texts is also more complex than Ray's division suggests. At first glance, it does seem that the earliest alchemical works do not feature many medical elements. For example, treatises such as the *Rasahṛdayatantra* or the *Rasārṇava* do not provide recipes for preparations specifically formulated to counter disease. By contrast, later texts devote large sections of their works to the therapeutic potential of mercurial and other formulations, broadly relying on ayurvedic categories and theories of disease causation. This development happens quite early on in the history of Indian alchemical literature: Medicine is already an important feature in the circa twelfth-century *Rasendracūḍāmaṇi*, which discusses the therapeutic actions of a series of substances used for alchemical processes.[2] The thirteenth- to fifteenth-century *Rasaratnākara*[3] dedicates one of its five sections to medicine: using ayurvedic disease categories, its *Rasendrakhaṇḍa* discusses various groups of diseases and gives recipes for medicines to treat them.[4] The *Rasaratnasamuccaya* devotes more than half its contents to the medical applications of its mercurial formulations, while the fifteenth-century *Rasendrasārasaṃgraha* eschews all non-medical applications of mercurials altogether.

and a later applied alchemy (*rasaśāstra*), medical alchemy (*rogavāda*), or mercurial medicine (*rasacikitsā*). Hellwig 2009a discusses Ray's chronological division and suggests other methods for deducing alchemical texts' chronological position.

[2] The *Rasendracūḍāmaṇi* discusses the medicinal properties of many of the substances it introduces in its tenth to fifteenth chapters. It does not, however, provide recipes for medicines directed against specific diseases.

[3] This work's five parts may date to different periods, with the parts dedicated to medicine probably belonging to a later strata. On the dating of the *Rasaratnākara*, see Meulenbeld 1999–2002: IIA, 664–665 and also Cantú, in this volume. Hellwig 2009a: 62–64 provides some evidence for different layers of the texts dating to different periods.

[4] Meulenbeld 1999–2002: IIA, 654–655, relying on a 1987 edition of the Śrīveṅkaṭeśvar Steam Press, lists seventy-one chapters. However, manuscript evidence suggests that the *Rasendrakhaṇḍa* only contains twenty-eight chapters. See Wujastyk 1984: 72 on the *Rasaratnākara* manuscripts held at the Wellcome Library in London and the addendum to this article at https://cikitsa.blogspot.com/2022/05/further-notes-on-rasendramangal.html. An unpublished study by Madhusudan Rimal of the digitized *Rasaratnākara* manuscripts held at the Berlin Staatsbibliothek confirms the shorter version of the *Rasendrakhaṇḍa*. On this subject, more research is needed, based on the available manuscripts of the *Rasendrakhaṇḍa*.

However, even the oldest alchemical works show knowledge of and concern with medicine and present health as fundamental to the pursuit of alchemy. In the *Rasahṛdayatantra*, the attainment of health and longevity is considered a prerequisite for achieving its stated goal of enlightenment: "With one exception, what is better than an ageless and deathless body that is the seat of all knowledge, that is the root of righteousness, prosperity, pleasure and liberation?" (1.27). "One who is worn out with ageing and subdued by afflictions such as cough, respiratory disease, etc. and whose mental and sensory functioning is impeded cannot achieve absorption (*samādhi*)" (1.29). The same work lists the medicinal properties of mercury in juxtaposition with its ability to bestow immortality and liberation: "Having thickened, it cures disease; bound, it gives liberation; well-killed (i.e., made into ash), it bestows immortality. What promotes compassion more than mercury?" (1.3).

Aside from the idea of alchemical elixirs or the substances used for making them having therapeutic potential, alchemy's deep reliance on ayurvedic thought is evident in one of the key aspects of alchemy, namely in the procedures associated with the intake of mercurial elixirs. Here, we can observe a regimen that is closely modelled upon ayurvedic precursors.[5] Elixir regimen (*rasāyana*) described in alchemical works typically begins with the preparation of the body of the person who will take the elixir. It involves inner cleansing through various forms of purging, followed by a period of rest with a mild diet. The intake of the elixir proper over a period of time follows. This echoes prescriptions for vitalization therapy (also *rasāyana*) that are well attested in ayurvedic textual sources, and already detailed in the earliest ayurvedic works, the *Carakasaṃhitā* and *Suśrutasaṃhitā*.[6] The parallels concern both the method of application as well as the substances used for the cleansing period, and the diet followed for the recovery period. And, just as in ayurvedic vitalization therapy, the elixir is used over a longer period of time. Differences lie in the name of *kṣetrīkaraṇa* ("preparation of the body") given in some alchemical sources for this preparation period, and in the actual elixir/tonic administered or taken.[7]

[5] This is discussed in detail in Chapter 1 in this volume.

[6] Both these works date to the early centuries CE. See Meulenbeld 1999–2002: IA, 105–115 and 333–352 for the dates of the *Carakasaṃhitā* and *Suśrutasaṃhitā*, respectively.

[7] For a more detailed analysis of commonalities and differences between ayurvedic and alchemical sources, as well as variations among alchemical texts, see Dagmar Wujastyk 2017.

In turn, medical texts soon started incorporating alchemical substances into their medicinal formulations, adopting both alchemical techniques for preparing them and the alchemical rationale of the need for purifying materials to make them fit for purpose. While the use of metals and minerals, including mercury, is attested in ayurvedic works before the composition of alchemical treatises, an increased use of these substances and reference to purification and other alchemical procedures are evident in ayurvedic works from about the eleventh century. With ayurvedic works incorporating alchemical procedures, and alchemical works focusing more and more on medicine, the boundaries between the disciplines become somewhat less clear over time. However, the division between the literary lineages usually remains visible: works that take their main inspiration from the classical ayurvedic works predominantly contain medical formulations based on herbal ingredients and animal products. Works based on the alchemical tradition primarily describe formulations containing mercury (*rasayoga*) and other inorganic materials.

While it is clear that both alchemy and Ayurveda took inspiration from each other, the medicine developed in alchemical contexts is framed somewhat differently, hinting at some competition between Ayurveda and iatrochemistry, the medicine developed in alchemical milieus. In its twenty-eighth chapter, the *Rasaratnasamuccaya* notes the superiority of mercury-based formulations. It states: "Mercury is superior to other drugs, because it is applied in low dosage, it does not elicit revulsion, and it quickly produces health."[8] Ayurvedic herbal formulations, by contrast, require higher doses, and their effects are seldom described as immediate. Herbal medicines also often have a strong taste, typically extremely bitter, which can make their intake challenging and impact appetite. Mercury and mercury-based pills, on the other hand, are typically without taste and therefore are unlikely to induce initial strong reactions from patients.

The sixteenth-century *Rasendrasārasaṃgraha* expands on the above verse, claiming that "(h)erbal medicine is prescribed by doctors in curable cases, but in incurable cases, mercurial medicine is given. It is therefore considered the best."[9] This idea is to some extent applied to the structure

[8] *Rasaratnasamuccaya* 28.1: *alpamātropayogitvādaruceraprasaṅgataḥ /kṣipramārogyadāyitvād auṣadhebhyo'dhiko rasaḥ* // The same verse is found in *Rasaratnākara, Rasakhaṇḍa* 1.15 (see the edition in the Digital Corpus of Sanskrit). The term "*aruci*" may also refer to loss of appetite rather than to revulsion to the taste of the medicines.

[9] *sādhyeṣu bheṣajaṃ sarvamīritaṃ tattvavedinā / asādhyeṣvapi dātavyo raso 'taḥ śreṣṭha ucyate//* *Rasendrasārasaṃgraha* 1.4–5.

of the medical chapters in alchemical works, which may prominently feature diseases considered particularly hard or even impossible to treat in ayurvedic works. The chapter of the *Rasaratnasamuccaya* translated here exemplifies this approach, as is discussed below. However, it should be noted that the less serious diseases are also extensively featured in the alchemical works, with both mercurials and other remedies offered for their treatment. This suggests that alchemical writers considered iatrochemistry a complete alternative to conventional Ayurveda, and a superior one at that. Conversely, we find that ayurvedic works that included iatrochemical formulations in their repertoire did not necessarily emphasize them as superior medicines, simply adding them to their lists of remedies.[10]

The *Rasaratnasamuccaya*

The *Rasaratnasamuccaya* by Vāgbhaṭa is considered the classic work of Indian alchemy. The treatise probably dates to the fifteenth or sixteenth century, though some place it as early as the thirteenth century.[11] It seems to have been in broad circulation and was widely cited by later works. The treatise itself is a compilation of older sources, with large parts based on the *Rasendracūḍāmaṇi* and other alchemical texts.[12] The *Rasaratnasamuccaya* dedicates more than half of its thirty chapters to medicine and shows an advanced stage of iatrochemical thought. Chapters 1 to 11 describe the fundamentals of alchemical practice: chapter 1 sets alchemy within a Śaiva setting, beginning with a verse praising Śiva, followed by an enumeration of twenty-seven alchemical masters (*rasasiddha*) and authors of alchemical works. It then focuses on mercury, the central ingredient of alchemical

[10] A typical strategy one can observe in post-eleventh-century ayurvedic works is that mercurials are featured at the end of chapter on a particular disease, following on from a series of herbal remedies that typically outnumber the mercurials. This is, for example, the typical structure in the chapters of the therapy section of the sixteenth-century ayurvedic treatise *Bhāvaprakāśa*. There are exceptions to this rule. For example, the eleventh-century ayurvedic treatise *Cakradatta* gives special prominence to an iatrochemical formulation it ascribes to the alchemist Nāgārjuna. In this case, it is an iron tonic, rather than a mercurial, but its alchemical provenance is explicitly stated.

[11] On the possible dates of the *Rasaratnasamuccaya*, see Meulenbeld 1999–2002: IIA, 670–671 and IIB, 681. Meulenbeld proposes that the treatise likely dates to the sixteenth century because of its mention of a plant otherwise first found in the sixteenth-century ayurvedic treatise *Bhāvaprakāśa*. The mention of opium (*ahiphena*) and cannabis (*vijayā*) suggests that the work is unlikely to be earlier than the thirteenth century.

[12] See Meulenbeld 1999–2002: IIA, 667 for some of the identified sources. These include the *Rasahṛdayatantra*, the *Rasārṇava*, and the *Rasaratnākara*. Notably, no source has been identified for chapters 12 to 30, the medical chapters of the *Rasaratnasamuccaya*.

practice, declaring it able to produce liberation in a living body (*jīvanmukti*). It also tells the story of mercury's origins, noting its varieties, and the need to process mercury before applying it. The following chapters are dedicated to other materials used in alchemical procedures, the initiation of the alchemical student, and the set-up of the laboratory and the various devices and instruments used. The chapters on the substances, which are largely derived from the *Rasendracūḍāmaṇi*, already include medical content, as each substance's medical effects are noted. However, the main medical part begins with chapter 12. Chapters 12 to 25 list over fifty disease categories, giving short analyses of their causation and symptoms and then discussing treatment more exhaustively.[13] Medical prescriptions broadly follow ayurvedic categorizations of disease and disease causation, without, however, directly quoting from ayurvedic sources currently known to us. The order in which the diseases are introduced resembles that of the ayurvedic treatises ascribed to another Vāgbhaṭa, the seventh-century *Aṣṭāṅgahṛdayasaṃhitā* and *Aṣṭāṅgasaṃgraha.*[14] The disease etiology given at the beginning of each chapter and in other verses introducing disease groups are not borrowed from the eighth-century *Mādhavanidāna*, and their source is unknown.[15] Chapter 26 is devoted to *rasāyana* therapy in the ayurvedic sense, that is, vitalization therapy rather than the alchemical elixir regimen. Chapter 27 is about virility therapy (*vājīkaraṇa*). Chapter 28 gives recipes for metallic formulations (*lohakalpa*), and chapter 29 discusses the medicinal use of poisons. The last chapter, chapter 30, returns to the subject of mercury, describing the alchemical procedure of "digestion" (*jāraṇa*), and listing recipes for eighty-six mercurial formulations (*rasakalpa*) for medical use. The majority of medicines in the *Rasaratnasamuccaya* are mineral-based formulations, very often mercury-sulfur compounds.

Abdominal Swelling—Dropsy (*Udararoga*)

Chapter 19, the chapter translated here, is dedicated to the ayurvedic category of diseases located or originating in the abdomen (*udararoga*) that

[13] Chapter 12, verses 1–8 list fifty-five disease groups as the subjects of chapters 12 to 25, but some chapters describe further disease groups. See the notes 140–145 in Meulenbeld 1999–2002: IIB, 679.

[14] See Meulenbeld 1999–2002: IIB, 679n139.

[15] See Meulenbeld 1999–2002: IIA, 667. This is remarkable in that most ayurvedic works written after the *Mādhavanidāna* incorporate the latter's order of disease groups and follow its categorizations of disease aetiology.

present with swelling of the soft tissues.[16] We have used the somewhat old-fashioned "dropsy" as a relatively unspecific term to represent this group of diseases. Modern ayurvedic practice understands this disease group as "diseases of the abdomen" or "ascites."[17] Chapter 19 differentiates between *udara* with swelling through water or humoral fluids and *udara* in which the abdominal swelling is caused by the humoral substance "wind" (*vāta*).[18] The chapter also discusses treatment for the related disease groups of anemia (*pāṇḍu*) and jaundice (*kāmala*).

The *udara* diseases are considered great diseases (*mahāgada, mahāroga*) in the ayurvedic tradition. These are diseases that are severe and hard to treat.[19] Indeed, for some of them, physicians are advised to not take them on, or to seek special consent from relatives if treatment is undertaken, because of the drastic measures that need to be taken and the low chances of success.[20] The *Rasaratnasamuccaya* places the *udara* diseases at the beginning of the chapter, with the more easily treatable anemia and jaundice following later.

Following the general pattern of the medical section, the chapter does not explain the symptoms and pathology in detail, focusing instead on the formulation of medicines. Of thirty-four formulations in this chapter, twenty-two are mercurials. Most of these contain both mercury and sulfur. The remaining formulations are either herbal or herbo-mineral. Some of the formulations also contain organic poisonous substances, such as snake poison, aconite, or arsenic. Very often, formulations are given names, such as

[16] This disease category is already discussed in the oldest ayurvedic works, the *Carakasaṃhitā* and *Suśrutasaṃhitā*.

[17] "Ascites" more typically describes a specific type of *udara*, namely *jalodara* (water-abdomen), but is sometimes used for the whole group, as *jalodara* is considered the general outcome of any of the types of *udara*. For an outline of how modern practice understands this disease group and the more specific terms used for the different types of *udara*, see Chandra et al. 2019. The authors note that "Udara roga denotes generalized distension or enlargement of abdomen of any etiology. Udara roga in Ayurveda is not only limited to ascitis, accumulation of fluid in the peritoneal cavity but also includes gaseous distension, hepato-spleenomegaly of varied etiology, intestinal obstruction and intestinal perforation."

[18] Humoral theory, i.e., the concept that three primary substances, wind (*vāta*), bile (*pitta*), and phlegm (*kapha/śleśman*), sustain the body when they are in the right location and in the right amount and cause disease when in the wrong location in the body and in the wrong amount, is one of the fundamental concepts of ayurvedic medicine.

[19] See, for example, *Suśrutasaṃhitā Sūtrasthāna* 33 on diseases that are untreatable, except through *rasāyana*, i.e., vitalization therapy, because of their complications. The *Suśrutasaṃhitā* counts the *udara* group of diseases among the eight severe diseases. The *Aṣṭāṅgahṛdayasaṃhitā Nidānasthāna* 12.45–46 notes that all kinds of *udara* are very difficult to treat even in their onset and that otherwise healthy people may be cured with great effort in the early stages of the disease. The *Rasaratnasamuccaya* itself also counts *udara* as a great disease in its list of great diseases in chapter 21, verse 1.

[20] *Aṣṭāṅgahṛdayasaṃhitā Cikitsāsthāna* 15.77–78 mentions getting the consent of relatives before giving the patient poison in what is likely a last-ditch attempt at saving the patient.

"The one who conquers death" (*mṛtyuñjayarasa*) or "The beauty of the three worlds" (*trailokyasundara*), sometimes with reference to other authorities' opinion on them. And all formulations are described as effective against a range of disorders, as well as against dropsy, anemia, or jaundice.

Despite defining *udara* as one of the great diseases (in chapter 21, verse 1), the *Rasaratnasamuccaya* does not mention the difficulty in treating *udara* diseases, but simply gives formulations to treat them. Out of fourteen formulations for *udara*, half contain mercury. One intervention is surgical (verse 22). The latter intervention is already found in similar form in the *Suśrutasaṃhitā*.[21] Ayurvedic works, which understand dropsy as caused by a fault in digestion, address it primarily through purgative therapies followed by a diet that facilitates digestion. There are also some herbal formulations, and in more desperate situations, the administration of poisons, cauterization (*agnikarman*), and surgery. The *Aṣṭāṅgahṛdayasaṃhitā* describes a particularly drastic surgical intervention in this context, in which the abdomen is cut open, the intestine is pulled out, and the obstructed part is removed. The wound is then closed by making big ants bite the edges and then removing their bodies: their jaws act as sutures.[22] The surgery described in the *Rasaratnasamuccaya* (and in the *Suśrutasaṃhitā*[23]), which involves making a small incision, inserting a pipe, and draining the fluid in the abdominal cavity with it, seems positively harmless by contrast.

The treatment of anemia (*pāṇḍu*) in the older ayurvedic works often consists of the application of medicated ghee and also of preparations that contain iron (iron, iron slag, and iron rust).[24] The *Rasaratnasamuccaya* also offers several iron-based medicines against anemia. However, most of its prescriptions for anemia contain mercury, again typically coupled with sulfur.

As for jaundice (*kāmalā*) and late-stage jaundice (*halīmaka*), the early ayurvedic works describe their treatment with herbal preparations of medicated ghee, typically with plants that address disturbances of the humour "bile" (*pitta*). Again, the *Rasaratnasamuccaya's* recipes emphasize the use of mercury and sulfur. Another addition in its recipes against both

[21] The surgical intervention occurs in the chapter on growths (*gulma*) in *Aṣṭāṅgahṛdayasaṃhitā Cikitsāsthāna* 14.18.

[22] This is described in *Aṣṭāṅgahṛdayasaṃhitā Cikitsāsthāna* 15.108–112.

[23] This surgery is also found in *Aṣṭāṅgahṛdayasaṃhitā Cikitsāsthāna* 15.113–117.

[24] See, e.g., the recipes in *Aṣṭāṅgahṛdayasaṃhitā Cikitsāsthāna* 16.9, 15–16, 20–22.

anemia and jaundice is the use of copper. Some of the ayurvedic recipes for these diseases feature gold or silver, but never copper.[25]

Finally, it bears mentioning that a number of the formulations described in the *Rasaratnasamuccaya* are still used in modern ayurvedic practice in India, including for the treatment of the *udara* category of diseases.

Notes on the Text and Translation

The translation of chapter 19 is based on the 1927 edition of the text by Āśubodha Vidyābhūṣaṇa and Nityabodha Vidyāratna. The Sanskrit text can be found on SARIT (https://sarit.indology.info/). This edition contains a commentary by the editors, the *Bodhinī*, which we have consulted for our translation.

Technical Sanskrit terms for diseases, plants, and so on are given in brackets after the translation. Measurements of weight, size, and volume are converted into metric units, with the Sanskrit term given in a footnote. The *Rasaratnasamuccaya* lays out two systems of measurements in its eleventh chapter. Its prescriptions in the other chapters follow the Māgadha system of measurements.

Chapter 19 is mostly an assemblage of recipes, interspersed with brief descriptions of the diseases addressed with the medicines. For convenience, we mark each recipe with a number and the formulation's name (if it is given) above the relevant verses.

[25] We checked the chapters on anemia and jaundice in the following ayurvedic treatises: *Carakasaṃhitā*, *Suśrutasaṃhitā*, *Aṣṭāṅgahṛdayasaṃhitā*, *Siddhayoga*, *Cakradatta*, *Vaṅgasenasaṃhitā*, and *Bhāvaprakāśa*. None of these had copper formulations.

Chapter 19

[Symptoms of ascites (*jalodara*)]

1. One whose abdomen is filled with fluid and humoral substances (*doṣa*) and has no folds, who has edema (*śvayathu*) or edema (*śopha*) of the legs: this may be a symptom of ascites (*jalodara*).

[Symptoms of "wind-dropsy" (*vātodara*)]

2. An abdomen filled with air and that is painful, emaciation, and especially shortness of breath: that is a symptom of "wind-dropsy" (*vātodara*).

Recipe 1

3. Powders[26] of mica, iron, mercury, sulfur, realgar (*śilā*), yellow arsenic (*āla*), copper (*tāmra*), the three pungent substances (dry ginger, long pepper, black pepper), leadwort (*citraka*), costus (*kuṣṭha*), musli (*muṣalī*), aconite(*viṣa*), and carom seeds (*dīpya*) mixed with lemon juice and made into pills: taken with honey at night, this cures all types of dropsy.

Recipe 2: "Bearer of happiness and knowledge" (*vinodavidyādhara*)

4. Mercury that is macerated well with equal amounts of sulfur, borax, purging croton will become the "Bearer of happiness and knowledge" (*vinodavidyādhara*).
   Administered with milk and jaggery, it removes all diseases that are treated with purgation (all types of) fever,[27] abdominal diseases, anal diseases and the concomitant pain in humans.
5. One should drink a decoction of split green gram after purgation (by the above medicine) has been induced. If there is severe diarrhea, one should take buttermilk and acacia bark juice.

Recipe 3

6. Eight parts of dehusked croton seeds, three parts of ginger, two parts of sulfur, and two of pepper, one part each of borax and mercury,

[26] The text uses the term "*cūrṇa*"-powder, which could just be the ground raw substance. In the case of metals and minerals, this powder is usually the calcined ash (*bhasman*) of the substances.

[27] The *Rasaratnasamuccaya* deals with fevers in its twelfth chapter. Following the categories of the early ayurvedic works, it differentiates between eight different kinds of fevers, caused by single humors, or by combinations of them. See *Rasaratnasamuccaya*, chapter 12, verses 9–12.

separately. 120mg[28] of this (mixture) is a very good purgative and should be administered together with cold water. It pacifies edema (*śopha*), tumors[29](*gulma*), and ascites (*jalodara*), and it is the best (medicine) for eliminating spleen diseases.

Recipe 4: "The conqueror of death" (*mṛtyuñjaya*)

7. The two types of alkali (*kṣāra*), the three pungent substances (dry ginger, pepper, and long pepper), five types of salts,[30] and fennel: equal amounts of the powder of these should be sieved through cloth.
8. One should make an equal amount of mercury and sulfur into a nice black mixture (*kajjalī*)[31] and mix everything together and triturate it for three days.
9. This medicine, which is called "The conqueror of death" (*mṛtyuñjaya*), acts very quickly. According to Malayārya,[32] it especially removes the mixed type of fever (*sannipātajvara*).[33]
10. For the mixed type of fever, this medicine should be taken in a dosage of 600 mg,[34] or laced with leadwort, dried ginger, rock salt, and the three pungent substances.
11. After drinking water, one who is afflicted by the three humors should stay in a windless area. Curd rice should be given as suitable food if requested and not otherwise. If its property does not manifest on him, the medicine should be given to him again.
12. It will remove all "wind" (*vāta*) diseases as well as "phlegm" (*kapha*) disorders, sluggish digestive power, fever, pain, all major diseases (*mahāmaya*), abdominal pain, liver diseases, anemia, edema (*śopha*), and pain due to visceral swelling, intestinal diseases, spleen diseases, constipation, vomiting due to visceral organ swelling, and dypsnea accompanied by cough, as well as hiccups.

[28] = one *guñja*.

[29] This can be tumors or visceral organ swelling.

[30] The five types of salts are rock salt (*saindhava*), sea salt (*samudra*), black salt (*biḍa/viḍa*), potassium nitrate (*sauvarcala*), and earthen salt (*romaka*).

[31] "*Kajjalī*" usually refers to black sulfide of mercury in alchemical texts, which has the form of a black powder.

[32] Different editions of the text give different readings for the name of this authority, who in this version seems to hail from the Malabar region.

[33] "Mixed type of fever" refers to the theory of its causation, i.e., that it is caused by a combination of vitiated humours. This is considered very difficult to treat.

[34] = five *raktikā*.

Recipe 5: Purgative (*virecana*)

13. It is said that first, a purgative should be given in all the types of dropsy.
    10 gm[35] of castor oil with cow urine or cow milk should be diligently administered. After cleansing,[36] the above medicine should be given.

Recipe 6: "Beautiful in the three worlds" (*trailokyasundara*)

14–16. Purified mercury and sulfur, calcined mica, rock salt, aconite, black cumin, and false black pepper, guduchi starch, leadwort, cardamom, barley alkali: each in half the amount of the mercury. One should triturate this with chaste tree juice for one day and with lemon juice for another day and then let it dry. This medicine is the "Beautiful in the three worlds" (*trailokyasundara*).
    240 mg[37] of this should be licked with ghee. This ends all types of "wind" dropsy (*vātodara*).

17. One should cook 48g[38] of leadwort root with two parts of cow urine and four parts of water until it becomes a paste (*kalka*). One should add ghee to the paste.
18. After adding 360g of barley alkali[39] and cooking it, one should remove it. One should take 10g with hot and greasy food.

Recipe 7: "Great fire" (*mahāvahni*)

19–21. One should mix together and finely grind four parts of mercury, eight parts of sulfur, two parts each of turmeric, the three myrobalans, and chebulic myrobalan, and eight parts each of black pepper, long pepper, ginger, cumin seeds, and wild croton. One should successively macerate [this] with river hemp (*jayantī*), common milk hedge (*snuh*), false daisy (*bhṛṅga*), leadwort, and castor oil, seven times each. One should take 4g[40] of this medicine, which is called "Great fire" (*mahāvahni*), with warm water.

[35] = one *karṣa*.
[36] i.e., after internal cleansing through purgation.
[37] = two *guñja*.
[38] = one *pala*.
[39] = one *valla*.
[40] = one *niṣka*.

22. This will induce purgation. One should eat buttermilk rice with rock salt at the end of the day as is suitable and avoid cold water. One should tap water above the umbilicus. This removes ascites.[41]

Recipe 8

23. 48g[42] of jaggery and dried ginger should be used for the removal of all types of dropsy.

Recipe 9: "Thunderbolt alkali" (*vajrakṣāra*)

24–25. One should macerate ground sea salt, rock salt, glass (salt of potash), barley alkali, black salt, borax, and sodium bicarbonate, all in equal parts, together with madar latex and common milk hedge's latex, and one should let it dry in the sun for three days. One should smear madar leaves with this mixture, enclose it, and apply a process of covered incineration.

26. This alkali should be pulverized and then mixed with equal quantities of dry ginger, black pepper, and long pepper, the three myrobalans, cumin seeds, turmeric, leadwort, and Java long pepper.

27. One should mix the alkali with half its amount of this mixture and administer it. 8g[43] should be eaten in cases of slow digestive function (*agnimāndya*) and indigestion (*ajīrṇa*).

28. When "wind" (*vāta*) is high, it is beneficial (to take it) with warm water, when "bile" (*pitta*) is high with ghee, in cases of (heightened) "phlegm" with cow urine, and when all three humors are disturbed with sour rice gruel.

29. This perfect "Thunderbolt alkali" (*vajrakṣāra*) is described by the bow-carrier (= Shiva). One should use in all the types of dropsy, visceral organ swelling (*gulma*) and pain caused by edema (*śopha*).

Recipe 10

30. One should drink 10g[44] each acacia, deodar with cow urine. This eradicates dropsy, anemia, pain, and spleen diseases.

[41] This surgical intervention is described more fully in *Suśrutasaṃhitā Cikitsāsthāna* 14.18.
[42] = one *pala*.
[43] = two *niṣka*.
[44] = one *karṣa*.

Recipe 11

31. One should macerate long pepper powder with milk hedge's latex for a day. 4g[45] (of the mixture) eradicates dropsy. One should drink it with buffalo urine.

Recipe 12: "Universal fire" (*vaiśvānara*)

32–34. One should prepare two parts each of mercury, sulfur, copper, shilajit, iron, the three pungent substances (dry ginger, long pepper, black pepper), leadwort, costus, five-leaved chaste tree, white musli, aconite, and carom seeds. Having ground everything, one should macerate it twenty-one times with a neem decoction and seven times with false daisy (juice).
(One should bind the mixture with) honey and administer it as a dried pill in the night. This formulation "Universal fire" (*vaiśvānara*) dries out ascites (*jalodara*).

Recipe 13: "Sun's radiance" (*sūryaprabhā*)

35–37. Blue glory tree (*bhārgī*), leadwort, the two kinds of sesban (*jayāyuga*), mica (*abhraka*), banana tree, velvet leaf (*pāṭhā*), sweet flag (*vacā*), cow gallstones (*rocanā*), Java long pepper (*cavya*), bay leaf, leadwort,[46] the three pungent substances, barley alkali and sodium bicarbonate, sulfur, fig plant (*trāyantī*),[47] quicksilver (*haravīja*),[48] Indian rose chestnut, aconite and snake poison, cloves, long pepper, costus root, frankincense (? *śallaphala*) together with the three myrobalans, also cuttlefish bone (*samudraphena*), Indian pennywort seed (*brāhmīvīja*), flame of the forest seeds (*latāvīja*), buds or sprouts of raw wood apple, the five types of salts, the eight types of flowers beginning with jasmine: these are mixed together with castor oil by the best physicians. This pill, known as "Sun's radiance" (*sūryaprabhā*) kindles the digestive fire.

[45] = one niṣka.

[46] Black *citraka* [PC].

[47] *Trayantī* is a synonym of *trāyamāṇa*. Other editions read sesban tree (*jayantī*) here.

[48] The presence of quicksilver is unusual here, given the absence of sulfur in the recipe, which mercury is usually coupled with in the *Rasaratnasamuccaya*'s recipes. This formulation is otherwise a completely herbal formulation with salt and poisonous content in it, so perhaps *haravīja* is a faulty reading here.

Recipe 14: "Rising sun" (*udayamārtaṇḍa*)

38. One should make thin sheets of 50g of copper. Then add an equal amount of sulfur and crush it all in a mortar.
39. Having added lemon juice to the mixture, one should place it in the sun for one day. Then, once the copper has become liquid, one should add 10g[49] of mercury.
40. Once it is properly prepared, it can be used for dropsy, edema, and fistula. This medicine, called "Rising sun" (*udayamārtaṇḍa*), is very famous.

[Symptoms of anemia (*pāṇḍu*)]

41. Discoloration of the body, swelling, leanness, loss of energy, and listlessness are the symptoms of anemia (*pāṇḍuroga*).

Recipe 15: "Swan iron rust" (*haṃsamaṇḍūraka*)

42–43. One should finely grind iron rust (*maṇḍūra*) and cook it in eight times its amount of cow urine. One should grind equal amounts of the three pungent herbs, the three myrobalans, Java grass, false black pepper, java long pepper, leadwort, Indian berberi, long pepper root, deodar. Having cooked equal amounts of ghee and iron rust, one should then mix it.

44. One should take 10g[50] (of this) and imbibe buttermilk after it has been digested. It is no surprise that this medicine, which is called "Swan iron rust" (*haṃsamaṇḍūraka*), removes anemia (*pāṇḍu*), late-stage jaundice (*halīmaka*), paralysis of the thigh (*ūrustambha*), jaundice (*kāmala*), and piles (*arśa*).

Recipe 16: "Destroyer of death" (*kālavidhvaṃsana*)

45. One should triturate equal parts of purified mercury, gold, silver, and copper mixed with citron juice in the sun for a day.
46. Having added the same quantity of mercury to the whole mixture, one should prepare a powder and cook it inside a thorn apple fruit in a "swing-device" (*dolāyantra*) for three days.
47. The device should be repeatedly refilled with datura juice. Having taken (the mixture) out of the swing device, one should wrap it in a cloth and cook it in a brick apparatus (*iṣṭikāyantra*).

[49] = one *karṣa*.
[50] = one *karṣa*.

48. Having made citrus juice and sulfur into a paste, one should place it above and below (the mixture). An equal amount is repeatedly added and, after it has been enclosed, cooked in a small furnace (*laghupuṭa*).
49. Once sulfur has been digested (i.e., amalgamated) six times, one should add an equal amount of calcined iron (*mṛtalohaka*) and triturate this for one day, and with the juice of yellow berried night shade (*kaṇṭakārī*) for one day.
50. Having enclosed it, one should roast it in a "pigeon furnace" (*kapoṭapuṭa*) with cow dung. Having ground it again, it should be roasted three times more with the same juice.
51. Similarly, one should triturate and roast it three times with the juice of the Indian nightshade and twice each with leadwort, madar, and Indian beech tree juice.
52. Having triturated, enclosed, and roasted it, one should also add a tenth part of aconite and grind it in and then administer 120mg[51] of this.
53. This medicine is called the "Destroyer of death" (*kālavidhvaṃsana*). It removes anemia diseases. One should administer chebulic myrobalan mixed with cow urine afterward.

Recipe 17: "The one with five faces" (*pañcānana*)

54–55. Ten g[52] each of calcined magnetic iron and gold, and copper, silver, and mica ash, all made into a fine powder that can pass through cloth and at once rubbed together with an equal quantity of a mercury-sulfur mixture (*kajjalī*): one should grind it with 120g[53] of chalcopyrite powder.

56. Having placed 96g[54] of orpiment powder into a crucible, and having added the mercury-sulfur mixture, one should add as much realgar.
57. Then, having covered it and having dried it thoroughly, one should roast it at night in an "elephant" pit furnace (*gajapuṭa*). Once it has cooled down on its own, one should grind it into a powder.

[51] = one *guñjā*.
[52] = one *akṣa*.
[53] = two and a half *pala*.
[54] = two *pala*.

58. One should add a mercury-sulfur mixture made with four parts of sulfur to the above mixture and triturate it with citron juice.
59. One should roast it ten times in boar pit furnace (*kroḍapuṭa*), and next, one should similarly roast it ten times with a mixture of mercury and arsenic.
60. And then, one should grind 1/16th part (of the above mixture) with calcined tourmaline (*vaikrānta*) ash and carefully pour it into a bottle.
61. This (medicine), called "The one with five faces" (*pañcānana*), is praised by the king of the gods. It is the best of all the mercurials, similar to the great substances (*mahārasa*)[55] in its properties.
62. It can be taken with chebulic myrobalan, yam, and dry ginger with ghee. It removes all types of anemia diseases like an ingrate removes all the good deeds.
63. It will remove wasting diseases, dropsy, late-stage jaundice (*halīmaka*), pain, "wind"-diseases (*vātārtti*), constipation, skin diseases (*kuṣṭha*), colitis, fever, diarrhea, dyspnea, cough, low appetite, "phlegm"-disease (*śleṣmavyādhi*), throat diseases, hemorrhoids, low digestion, urinary disease, pain related to swelling diseases, and whatever other disease there is.
64. While taking this medicine one should avoid only bael. A healthy person should eat everything together, and an ill person should eat what is wholesome and counters the disease.

Recipe 18: "Ocean of health" (*ārogyasāgara*)

65. A mercury-sulfur mixture (*kajjalī*) consisting of 48g[56] each of sulfur and mercury, with 96g[57] of copper pyrite and arsenic;
66. Having added 48g of realgar, 48g of mica ash, and 10g[58] of borax (*sukhasparśa*), one should grind it.
67. One should place (the above mixture) in a covered crucible, sandwiched between 30g of purified and clean copper sheets.
68. Having tightly sealed the crucible with mud and cloth and having let it dry, one should roast it with cow pats in an elephant furnace.

[55] The term *mahārasa* refers to a specific group of mineral substances, such as copper sulfate, mica, and copper pyrite. See Hellwig 2009b: 227–228. The *Rasaratnasamuccaya* discusses the *mahārasa* at some length in its second chapter.

[56] = one *pala*.

[57] = two *pala*.

[58] = one *karṣa*.

69–70. Once it has cooled on its own, one should lift out the (mercury) that has become "lame" (*khoṭībhūta*) and pulverize it. One should roast the mortar-ground mixture together with the powders of sulfur, arsenic, and realgar ten times in a boar furnace. And next, one should add tourmaline ash with twenty parts of the above mixture.

71. Having ground it, one should make a bolus and place it in silver bowl. This most beneficial medicine is called the "Ocean of health" (*ārogyasāgara*).

72. 240g[59] of it with pepper and ghee removes anemia, low appetite, anal diseases, diseases of "wind," "bile," and "phlegm," growths, bloating, edema, dyspnea, headaches, vomiting, very low digestive power, "heavy retention,"[60] various fevers, and other diseases in the context of pregnancy or birthing.

Recipe 19

73. One should cook equal parts of copper ash, mercury ash, sulfur, and aconite that have been triturated with leadwort juice for forty-five minutes[61] in mild heat.

74. 240g of it will remove edema and anemia like drying up mud.

Recipe 20

75. One should take 336g[62] each of finely ground porcupine flower plant, false daisy, asparagus, and spreading hogweed (*punarnava*).

76. In a decoction of the above, one should properly cook more than 360 chebulic myrobalan fruits. After drying, one should let it cook in cow's milk.

77–78. Having dried it slowly and having removed (the seeds?), one should fill (them) with pills. One should take 144g of mercury and cook it in 144g of sulfur and then press it into a leaf and grind it again. Then one should add 384g[63] of dried guduchi.

79. Having ground everything together with honey, one should make small pills.[64] One should string them together and place it in pot of

[59] = two *rati*.
[60] The disease *udāvartta* is understood as a change in the direction of the flow of the humor "wind."
[61] = 1/4 *yāma*, one *yāma* = 3 hours.
[62] = seven *pala*.
[63] = eight *pala*.
[64] These should be placed into the previously prepared hollowed myrobalans according to the commentary.

honey. One should always take these one by one (i.e., one a day) to eradicate dry anemia.

Recipe 21

80–81. One should take four parts of mercury, two parts of the described iron, many parts of leadwort, nut grass, false black pepper, the three pungent substances (dry ginger, pepper, and long pepper), the three myrobalans, and also tellicherry bark (*kuṭaja*). Having ground it all, one should make into pills with honey. One should take one pill at a time in the morning in order to remove anemia caused by the humour "bile" (*pittapāṇḍu*).

Recipe 22: "Beautiful of the three worlds" (*trailokyasundara*)

82. One should soak mercury, sulfur, iron, mica, guduchi starch, air potato (*sūkara*), the three myrobalans, and moringa roots with false daisy juice.
83. This "Beautiful of the three worlds" (*trailokyasundara*) mixed with ghee, honey, and sugar removes anemia and edema in one who eats what is wholesome like the moon.

Recipe 23: "Three-eyed" (*trinetra*)

84. One should take the medicine known as the "Three-eyed" (*trinetra*) mixed with some ghee, honey, jaggery, quail (meat), and guggul with water.[65]

Recipe 24: "The victorious" (*vijayā*)

85. 144g of chebulic myrobalan, 144g of leadwort, 24g each of cardamom, cinnamon, bay leaf, nut grass is suitable;
86. 24 g of chaste seed as indicated, and half its amount of cobra saffron, 48g of the three pungent substances, long pepper root, and aconite;

[65] A recipe for a *trinetrarasa* occurs in *Śārṅgadharasaṃhitā* 2.12, 216 and is repeated in *Rasamañjarī* 6.216 (see the Digital Corpus of Sanskrit for these passages). It is tempting to think that the recipe for *trailokyarasa* in verse 81 here might actually be the recipe for *trinetrarasa*, as there is already a different recipe for a *trailokyarasa* in this chapter in verses 14–18. However, the recipe in verse 81 is different from the recipe in the *Śārṅgadharasaṃhitā* and *Rasamañjarī*.

87. 48g of mercury and of sulfur: one should grind these into fine powders. One should add this to 2500g[66] of melted old jaggery.
88. A wise person should take it when it is cool to the touch, having applied ghee to the hands. The size of the seed of a jujube (*badara*), the pill is known as "The victorious" (*vijayā*). One should let (a patient) eat this (pill), which removes anemia and edema at night.
89. One should eat borax and amaranth to counteract (the pill's) toxins.[67]

Recipe 25

90. One should mix equal amounts of mercury, sulfur, copper ash, croton, and guggulu with ghee and make this into a pill. The doctor should administer the preparation one at a time in order to remove edema and anemia.

Recipe 26

91. One should regularly take 4g[68] of the powder of the five parts of the bristly luffa (*devadālī*) with milk or water for a month to remove anemia.

[Symptoms of jaundice (*kāmalā*)]

92. Jaundice presents with burning flesh and blood and "bile" (*pitta*) in the blood, tiredness, thirst and a burning sensation. Through neglect and in connection with edema it becomes late-stage jaundice (*kumbhakāmalā*), which presents with yellow eyes and yellow stools.

Recipe 27

93. One should crush prickly chaff and khejri tree roots and let (the patient) drink it with buttermilk. 10g[69] of it removes jaundice, edema, and anemia.

Recipe 28

94–95. Wrought iron (*tīkṣṇa*), copper pyrite, magnetic iron, mica, copper, mercury, and arsenic, triturated with bristly luffa juice

[66] = half a *tula*.
[67] We count this verse as part of the recipe for "The victorious" pill, with borax and amaranth as antidotes to the specific toxicity of the pill's ingredients.
[68] = one *niṣka*.
[69] = one *karṣa*.

and thickened in a sand apparatus (*vālukāyantra*) and mixed with guduchi, lotus bulbs, *kahular* bulbs, and black raisins and ground with licorice juice, honey, and sugar removes Jaundice.

Recipe 29: "Of three sources" (*triyoni*)

96. One should smear a quarter of copper with mercury that has risen. Having mixed it with lemon juice, one should expose it to the sun.

97–98. Having placed sulfur above and below, one should surround it with sessile joyweed (*matsyākṣī*), seal it with clay, and let it bake with great care. After it has been baked for eight hours[70] and has cooled down, one should take it out.

One should give a dosage of 120mg[71] of this with chebulic myrobalan and jaggery.

This medicine is known as the "Of three sources" (*triyoni*). It removes anemia and edema.

Recipe 30

99. Nut grass, guduchi, leadwort, licorice, long pepper, false pepper, dry ginger, and the three myrobalans, ground into a powder one by one, mixed with iron powder and copper pyrite, is the best cure for anemia.

Recipe 31: "Lord of love" (*kāmeśvara*)

100. One should mix 48g of mercury, 48 g of sulfur, 144g each of common milk hedge and chebulic myrobalan, and 24g each of nut grass, cardamom, and bay leaf;[72]

101. 48g each of the three pungents, long pepper root, and aconite, 10g of cobra saffron, 24g of chaste seed.[73]

102. One should cook this with 2500g[74] of melted old jaggery and should triturate it for three hours with aloe vera juice and with ghee.

103. One should make a pill like a jujube seed and administer it at night. This medicine, which removes edema and anemia, is itself the "Lord of love" (*kāmeśvara*).

[70] = one *yāma*.
[71] = one *guñjā*.
[72] 48g = one *pala*, 144g = three *pala*, 24g = half a *pala*.
[73] 48g = one *pala*, 10g = one *karṣa*, 24g = half a *pala*.
[74] = half a *tula*.

Recipe 32

104. One should cook equal parts of copper ash, mercury ash, sulfur, and aconite triturated with leadwort juice, for forty-five minutes[75] on mild heat.

105. An amount of 240g removes the wind diseases, edema, and anemia like drying mud.[76]

Recipe 33

106. Mercury that is ground and cooked with bronze and realgar and mixed with wrought iron (*tīkṣṇa*) ash and copper ash removes late-stage jaundice (*halīmaka*).

Recipe 34

107. One should triturate 50g of purified mercury and 50g of vermillion (*sindūra*) with malabar nut juice for three hours and form a disc.

108. One should cook it well in a crucible twelve fingers high. One should add 192g[77] of purified sulfur to it.

109. Having made the above-mentioned disc and having added it, one should bake it for a short time. Once the sulfur has been absorbed, one should take the disc out and pulverize it.

110. One should add ten times the amount of iron ash to the powder and triturate it with a tenth amount of garlic. One should prepare pills in the size of horse gram.

111. This medicine, the "Vermillion ornament" (*sindūrabhūṣaṇa*), is prepared as a cure for "wind"-induced anemia. After its administration, one should take prickly chaff flower and castor root crushed in buttermilk. 10g[78] remove anemia and jaundice.

Thus, this is the nineteenth chapter, on the treatment of dropsy, anemia, edema, jaundice, late-stage jaundice, and end-stage jaundice in the *Rasaratnasamuccaya* by the learned Vāgbhaṭa, son of the master physician Siṃhagupta.

[75] = a quarter of a *yāma*.

[76] Note by P. Chorge: This medicine absorbs or removes the "wind" (*vāta*) diseases, edema, and anemia similar to the way the sun absorbs or removes water from the mud. See also verses 73–74.

[77] = four *pala*.

[78] = one *karṣa*.

Bibliography

Ācārya, Yādavaśarman Trivikrama. 1913. *Śrīnityanāthasiddhaviracito Rasaratnākarāntargataś caturtho Rasāyanakhaṇḍaḥ . . . saṃśodhitaḥ prakāśitaś ca = Rasayanakhanda, Fourth Part of Rasaratnâkara by Shri Nityanâthasiddha edited by Jadavjî Tricumjî Âchârya.* Āyurvedīya Granthamālā 10. Bombay: Nirnayasagar Press.

Ācārya, Yādavaśarman Trivikrama. 1927. *Rasahṛdayatantram. Śrīmadgovindabhagavatpādav iracitaṃ. Caturbhujamiśraviracita-Mugdhāvabodhinīvyākhyāsamullasitaṃ. Kāḷe ityupāhv agurunāthātmajatryambakena tathā Ācāryopāhvena Trivikramātmajena Yādavaśarmaṇā saṃśodhitam.* Lahore: Motilal Banarsidass/The Bombay Sanskrit Press.

Ācārya, Yādavaśarman Trivikrama, ed. 1981. *Carakasaṃhitā, śrīcakrapāṇidattaviracitayā āyu rvedadīpikāvyākhyayā saṃvalitā.* 4th ed. New Delhi: Munshiram Manoharlal.

Aṣṭāṅgahṛdayasaṃhitā. See Srikantha Murthy 1999–2000.

Aṣṭāṅgasaṃgraha. See Srikantha Murthy 1995–1997.

Bhāvaprakāśa. See Srikantha Murthy 1998–2000.

Cakradatta. See Sharma 2002.

Carakasaṃhitā. See Ācārya 1981.

Chandra, Nikhil, et al. 2019. "Critical Analysis of Udara Roga with Special Reference to Contemporary Science." *World Journal of Pharmaceutical and Medical Research* 5 (10): 156–159. ISSN 2455-3301.

Hellwig, Oliver. 1999–. *DCS: Digital Corpus of Sanskrit.* url: http://www.sanskrit-linguistics.org/dcs/

Hellwig, Oliver. 2009a. "Alchemical Procedures and Their Implications for the Chronology of Medieval Rasaśāstra." *eJournal of Indian Medicine* 2: 53–65.

Hellwig, Oliver. 2009b. *Wörterbuch der mittelalterlichen indischen Alchemie.* Groningen: Barkhuis & University of Groningen, University Library.

Meulenbeld, Gerrit Jan. 1999–2002. *A History of Indian Medical Literature.* 5 vols. Groningen: E. Forsten.

Miśra, Siddhinandan. 1984. *ācārya somadeva kṛta rasendracūḍāmaṇiḥ, "siddhipradā" hindivyākhyāsahitaḥ. Hindīvyākyākāraḥ Dr. Siddhinandana Miśra.* Vārāṇasi/ Dillī: Chaukhamba Orientalia.

Nadkarni, A. K. 1954. *Dr. K. M. Nadkarni's Indian Materia Medica, with Ayurvedic, Unani-tibbi, Siddha, Allopathic, Homeopathic, Naturopathic and Home Remedies, Appendices and Notes.* 2 vols. Bombay: Popular Prakashan.

Prabhakara Rao, G. 2015. *Rasendra Sāra Saṅgraha of Sri Gopala Krishna Bhatt. Sanskrit Text with English Translation.* Varanaṣi: Chaukhamba Orientalia

Rasahṛdayatantra. See Ācārya 1927; Subbarayappa 1997; and Rasaśāstri 1989.

Rasaratnākara Rasāyanakhaṇḍa. See Ācārya 1913.

Rasaratnasamuccaya. See Vidyābhūṣana 1927.

Rasārṇava. See Ray 2007.

Rasaśāstri, Daulatarāma. 1989. *Rasahṛdayatantram. Śrīmadgovindabhagavatpādaviracitaṃ. Caturbhujamiśraracita "Mugdhāvabodhinī" Saṃskṛta ṭīkā tathā Hindī vyākhyā sahitam, vyakhyākṛtṛ Daulatarāma Rasaśāstrī.* Vārāṇasī: Caukhambhā Oriyanṭāliyā

Rasendracūḍāmaṇi. See Miśra 1984.

Rasendrasārasaṃgraha. See Prabhakara Rao 2015.

Ray, Prafulla Chandra. 1902. *A History of Hindu Chemistry from the Earliest Times to the Middle of the Sixteenth Century A.D. with Sanskrit Texts, Variants, Translation and Illustrations.* London and Oxford: Williams and Norgate.

Ray, P. C., and H. C. Kaviratna. 2007. *Rasārṇava. Edited by P. C. Ray, H. C. Kaviratna. Re-edited by S. Jain.* Delhi: Oriental Book Centre.

Śārṅgadharasaṃhitā. See Srikantha Murthy 2006.

Sharma, Priya Vrat. 2002. *Cakradatta (Sanskrit Text with English Translation): A Treatise on Principles and Practices of Ayurvedic Medicine.* Varanasi: Chaukhamba Publishers.

Siddhayoga. See Tewari 2006.

Srikantha Murthy, K. R. 1995–1997. *Aṣṭāṅga Saṃgraha of Vāgbhaṭa, Text, English translation, Notes, Indices etc., translated by K.R. Srikantha Murthy*. 3 vols. Varanasi: Chaukhamba Orientalia.

Srikantha Murthy, K. R. 1998–2000. *Bhāvaprakāśa of Bhāvamiśra, Text, English Translation, Notes, Appendeces [sic] and Index, Translated by K. R. Srikantha Murthy*. 2 vols. Varanasi: Krishnadas Academy.

Srikantha Murthy, K. R. 1999–2000. *Vāgbhaṭa's Aṣṭāṅga Hṛdayam, Text, English translation, Notes, Appendix and Indices, translated by K.R. Srikantha Murthy*. 3 vols. Varanasi: Krishnadas Academy.

Srikantha Murthy, K. R. 2006. *Śārṅgadhar-Saṃhitā (A Treatise on Āyurveda) by Śārṅgadhara. Translated into English by Āyurveda-Vidwān Prof. K. R. Srikantha Murthy*. Varanasi and Delhi: Chaukhamba Orientalia.

Tewari, Premvati, ed. 2006. *The First Treatise of Āyurveda on Treatment Vṛndamādhava or Siddha Yoga*. Varanasi: Chaukhambha Visvabharati.

Vidyābhūṣaṇa, Āśubodha, and Nityabodha Vidyāratna, eds. 1927. *Saṭīkaḥ Rasaratnasamuccayaḥ: (prācīnarasagranthaḥ). mahāmati śrīmad vāgbhaṭācāryyaviracitaḥ. paṇḍita—śrīmad āśubodhavidyābhūṣaṇena tathā paṇḍita -śrīman nityabodhavidyāratnena ca viracitayā rasaratnasamuccayabodhinyākhyayā ṭīkayā samalaṅkṛtaḥ prakāśitaś ca*. Kalikātā: The editors.

White, David G. 2013. "Rasāyana." *Brill Encyclopedia of Hinduism* 5: 489–499.

Wujastyk, Dagmar. 2017. "Acts of Improvement." *History of Science in South Asia* 5 (2): 1–36. https://doi.org/10.18732/hssa.v5i2.26.

Wujastyk, Dominik. 1984. "An Alchemical Ghost: The Rasaratnâkara by Nâgârjuna." Ambix 31 (2): 70–83.

Wujastyk, Dominik. 2022. "Further Notes on the Rasendramaṅgala and Rasaratnākara." https://cikitsa.blogspot.com/2022/05/further-notes-on-rasendramangal.html.

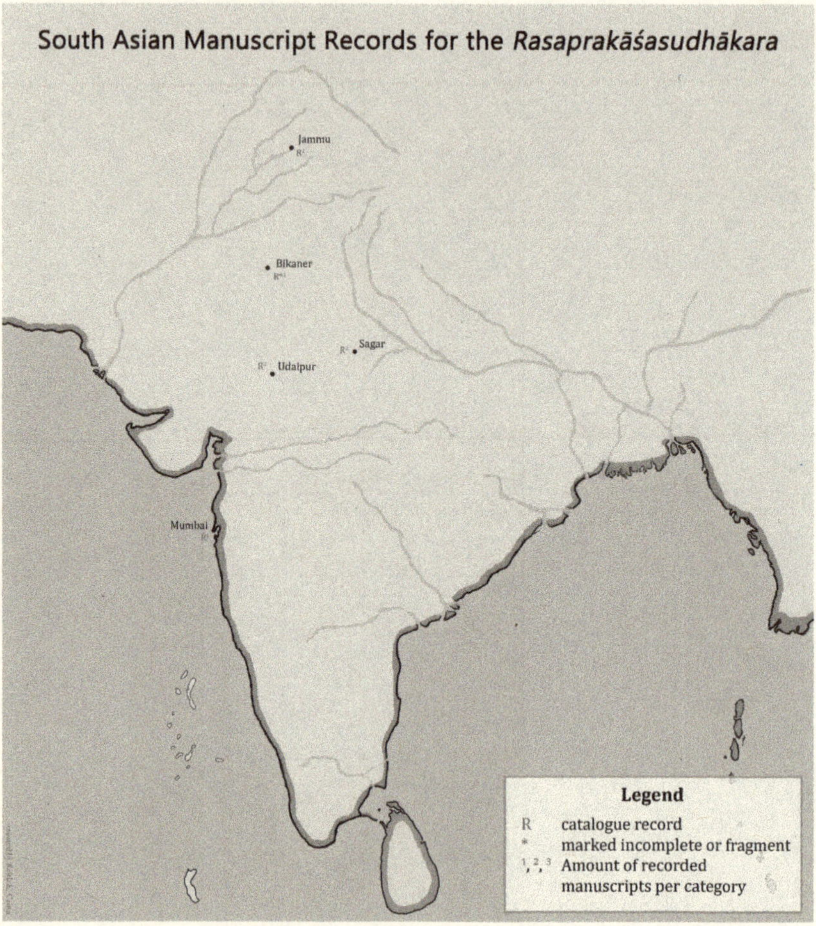

Map 8.1 South Asian manuscript records for the *Rasaprakāśasudhākara*.

8

The Eighteen Alchemical Procedures

Chapter 1 of the *Rasaprakāśasudhākara*

Dagmar Wujastyk

With all their differences in focus and aims, there is a widely shared feature among Sanskrit alchemical works: large portions of the texts, indeed, usually the better part of them, are dedicated to the technical aspects of alchemical practice. Philosophical speculation is rare, and religious explanation limited. Instead, practical instructions on how to produce desired outcomes make up the bulk of alchemical writing. The explanation of "how to?" trumps the question of "why does it work?" Most Sanskrit alchemical works delineate more or less systematic programs of alchemical operations, with lists and specific sequences of procedures for processing substances. Mercury is at the center of the alchemical endeavor, and thus, most of the alchemical procedures in the treatises are dedicated to its treatment. However, other substances also play important roles, either as a means for cleansing or enhancing mercury or as a product in their own right. The alchemical works' instructions describe how to use materials in relation to each other, how to achieve different aggregates through the application of heat or cold, and which instruments to use for each operation. They may also include detailed descriptions of ritual actions that precede or accompany alchemical operations, and the necessary setup for them.[1] Here, there is a divide between the works, some of which place great emphasis on ritual, while others pay little or no attention to it. This divide is already present in the oldest of the alchemical works transmitted to us, the tenth-century *Rasahṛdayatantra*, and the twelfth-century *Rasārṇava*. The latter shows great influence from tantric thought and lays out a ritual framework for alchemical practice that is reiterated in many later alchemical works, while the former does not even

[1] See, for example, Chapter 2 by David G. White and Chapter 4 by P. Sauthoff in this volume on the setup of the alchemical laboratory.

Dagmar Wujastyk, *The Eighteen Alchemical Procedures* In: *Indian Alchemy*. Edited by: Dagmar Wujastyk, Oxford University Press. © Oxford University Press 2025. DOI: 10.1093/9780197805831.003.0009

Figure 8.1 Mercury pursuing maiden on a horse. Mughal Painting, ca. 1750, entitled "Vark-a-Para," accession number AG 724. Image reproduced with kind permission of the Trustees of the Maharaja Sawai Man Singh II Museum, Jaipur.

provide passing reference to ritual or mantra, though it does refer to deities and to transcendent aims. This is thus not a divide between worldly or religious aims, but simply one of method. In this chapter, we will examine the program of alchemical operations laid out in the *Rasaprakāśasudhākara* and explore how its version compares with that of the other works.

The *Rasaprakāśasudhākara* (*The Nectar Mine Light on Mercury*) by Yaśodhara probably dates to the sixteenth century CE. It comprises 1286 verses in thirteen chapters that describe both alchemical procedures for making (or faking) precious metals and gems and the making and application of iatrochemical medicines.[2]

The work's first chapter lays out its alchemical program. It begins with a narrative about the method for acquiring mercury, the central substance of the alchemical endeavor. This is a story involving a maiden on a horse whose presence entices mercury to come out of its well and pursue her to subsequently be scattered in a large area around its original well. Variations of this

[2] For an overview and summary of the *Rasaprakāśasudhākara*, including its date, see Meulenbeld 1999–2002, IIA: 645–650.

story are found in several of the Sanskrit alchemical works, but intriguingly, also in sources outside Indic literatures.[3]

In the *Rasaprakāśasudhākara*, the story provides an explanation for the different locations in which mercury is found. The text ascribes different attributes to the mercury in each location and associates each of the four types of mercury with a different societal class. However, the significance of this association with different classes is not elucidated further. The types of mercury are also distinguished by their color: white, black, yellow, and red, and each color is attributed with a specific efficacy. The white kind is declared suitable for making silver, the yellow kind for making gold, the black kind for enhancing the body and the red one for eradicating disease. Mixed together, they eradicate disease and bestow special powers. While mercury is defined as a substance with desirable properties, it is also described as being endowed with several innate flaws and contaminants.[4] Therefore, mercury must go through a series of alchemical operations that will render it fit for purpose and cleansed of impurities and negative characteristics. Its innate good properties will be enhanced and its power heightened, until it finally attains the right condition needed for its ultimate purpose as an agent for transmuting base materials into noble ones and conferring health and longevity on its consumer.

Following a sequence first laid out in the tenth-century *Rasahṛdayatantra*, the *Rasaprakāśasudhākara* then delineates a systematic program of eighteen alchemical operations, called *saṃskāra*:[5] Sixteen for making the elixir, and a further two sets of procedures for the final processes of applying the elixir to metals and to the human body.

These are

1. steaming (*svedana*),
2. trituration (*mardana*),
3. thickening (*mūrchana*),

[3] See Sarma and Sahai 1995 for a discussion of the theme and the artwork depicting it. White 1996: 449 gives references to several Sanskrit alchemical works that contain variations of the story. White 2017 discusses the motif in other non-Indic literatures, notably in Chinese and Syriac sources.

[4] An origin myth given in the *Rasendracūḍāmaṇi* provides an account for mercury's flaws. While mercury is explained to be Shiva's semen in this story and therefore the most powerful substance possible, the god Indra asked Shiva to contaminate the mercury that had fallen to earth so that humans could not access its powers easily and therefore become equal to the gods. See *Rasendracūḍāmani* 15.21–22, and White 1996: 192.

[5] The term *saṃskāra* is probably the most ubiquitous among alchemical works to denote alchemical operations for making mercurial elixirs. However, some works use the term *rasakarman* instead. See also Chapter 1 in this volume on the *Rasahṛdayatantra*'s list of *rasakarman*.

4. risen (*utthita*),
5. causing to fall, distillation (evaporation and condensation[6]) (*pātana*),
6. reviving (*nirodhana*),
7. fixing (*niyamana*),
8. stimulating (*dīpana*),
9. measuring the morsel of mica (*abhrakagrāsamāṇa*) and
10. feeding it to the mercury (*cāraṇa*),
11. enclosed liquefaction (*garbhadruti*),
12. external liquefaction (*bāhyadruti*),
13. digestion, i.e., amalgamation (*jāraṇa*),
14. coloring, dyeing (*raṃjana*),
15. setting in motion, or potentiation (*sāraṇa*),
16. advancing the penetrating power (*krāmaṇa*),
17. transmutation (*vedha*), and finally
18. ingestion (*sevana*).

Each step represents a transition from one state to another: from contaminated to clean; from liquid to solid state or vice versa; from ineffective to powerful; and from dangerous to fit for purpose. This perhaps links the alchemical *saṃskāṛa* to their older namesake, the *saṃskāra* as rites of passage, which both mark transitions in life, and prepare a person for them in a successive process of purification.[7]

To this end, the main substance mercury is mixed and ground with other organic and inorganic materials, dipped in liquids or coated with pastes, and heated and cooled in different ways. Each step prepares the ground for the next. The boundaries between the various procedures are, however, somewhat fluid: it can be difficult to decide when one process concludes and another begins, and some of the procedures are very similar to each other.

[6] Hellwig 2009: 285 notes that *pātana* generally describes a process in which a reagent moves inside an apparatus. In the case of mercury, this is typically its evaporation and then condensation. In the case of other substances, this can mean the chemical separation of a substance in which its different elements deposit in different parts of the apparatus. Bhagwan Dash 2003: 63 translates *pātana* as "distillation" or "sublimation." Sublimation describes the transition from a solid state of matter to a gaseous state of matter, and thus covers only part of the *pātana* process. Distillation, by contrast, describes the conversion of a liquid into vapor that is subsequently condensed back to liquid form. Dry distillation, i.e., the heating of solid materials to produce gaseous products that may condense into liquids or solids, is perhaps the closest correlate. However, for simplicity's sake, I will refer to *pātana* as "distillation" here.

[7] See Michaels 2017 on the term *saṃskāra* and its history, and the main sources, the Gṛhyasūtras and Dharma texts.

What differentiates them is the stage in which they are applied in the overall process, which results in differently interpreted outcomes.

Alchemical Terms: Metaphors and Links with Ayurveda

The technical terms used for each operation are often descriptive of visible actions and changes in the mercury. Some bear close relationship to terminology used in ayurvedic medicine. Mercury is described in terms of a living, perhaps sentient, entity that is quasi-medically treated. One of the parallels with ayurvedic treatment is the application of "steaming" (*svedana*), one of several procedures to ready the patient for treatment proper. In ayurvedic medicine, patients are subjected to steam to make them sweat. This serves to cleanse them from bodily waste. In alchemical terms, *svedana* describes the application of steam or a boiling liquid to mercury for the purpose of cleansing it from contaminants. The term "stimulating" (*dīpana*) represents another metaphor with links to ayurvedic terminology. In ayurvedic terms, *dīpana* is the kindling of the digestive fire, enhancing a person's appetite and ability to digest foods. In alchemical terms, *dīpana* serves to stimulate the ability of mercury to "swallow" (*cāraṇa*) and then "digest" (*jāraṇa*) other substances, that is, amalgamate with them.

Some terms for alchemical operations are purely descriptive, as for example in the case of "trituration" (*mardana*), where mercury is rubbed with other substances. The names of procedures can also indicate the physical appearance or the presumed condition of the mercury in terms of its potency at the respective stage of processing. For example, in the procedure called "thickening" (*mūrchana*, in other texts also called *mūrchā*), the mercury visibly thickens, losing its defining characteristic of volatile mobility. Thus, *mūrchana* is a process of thickening, solidification, or coagulation. At the same time, the mercury loses its shine during this procedure, turning a dull, lustreless grey. In medical terms, *mūrchā* denotes "stupor," a condition associated with the loss of strength. And indeed, one of the presumed effects of the alchemical procedure is the weakening of the potency of the mercury, made visible in its loss of lustre. The mercury becomes of "low power" (*mandavīrya*) and "impotent" (*ṣaṇḍha*). This then necessitates raising (*utthāpana*), condensing (*pātana*) and reviving (*rodhana*) it, which will re-empower it and bring back both its mobility and shine. Thus, both a descriptive "thickening" and a metaphorical "stupor" (or, as White 1996: 266 has it, "swooning") represent important aspects of the procedure of *mūrchana*.

Categories of Substances

With the *rasakarman* or *saṃskāra*, Indian alchemists introduced a new concept of how materials function, namely, that substances—and mercury in particular—can be effective and powerful medicines or elixirs, but that they need to be cleansed and altered first. One of the innovations of Indian alchemy here is that the way materials are made fit for purpose involves treatment with other materials. That is, some materials act as a kind of medicine for others. Plant and animal substances, but also metallic and non-metallic minerals, are used to remove contaminants or to counteract a material's unwanted characteristics.

This concept gave rise to a new classification scheme of substances in alchemical works. In the early ayurvedic treatises, substances are generally categorized according to their visible characteristics and medicinal action and are grouped together based on their main applications.[8] In alchemical literature, the same substances are primarily classified according to their role in the alchemical process, though their therapeutic application may also be mentioned. Some materials, such as gold, silver, or some gems, are described both in terms of their use in processing mercury and as the final products of alchemical operations.

The *Rasaprakāśasudhākara* differentiates between metals (*loha*, *dhātu*), and two broad groups of what would now be considered minerals (*mahārasa* and *uparasa*).[9] It also describes various organic materials: herbal drugs for mercurial processing that are classified as "divine drugs" (*divyauṣadhi*), "essence drugs" (*rasauṣadhi*), "great drugs" (*mahauṣadhi*), and "drugs for success"[10] (*siddhauṣadhi*).

[8] The characteristics are their taste (*rasa*), post-digestive taste (*vipāka*), potency (*vīrya*), and specific action (*prabhāva*), and these directly act on a patient's digestive functioning and humoral balance. On ayurvedic pharmacological concepts, see Meulenbeld 2000.

[9] See Hellwig 2014 for a more detailed analysis of categories of inorganic materials in the Sanskrit alchemical sources. Also see White 1996: 446n18, where he notes that the "*mahārasas* act directly on mercury; the *uparasas* catalyze the effects of the *mahārasas*."

[10] The "success" being success in alchemical operations, though the drugs are also understood to bestow perfection of the body (*dehasiddhi*) and perfection of metals (*lohasiddhi*) by themselves.

Divine herbs and great herbs first find mention in the early ayurvedic works in the context of vitalization therapy (*rasāyana*), though the lists of plants do not entirely correspond with those of the alchemical works.[11] In the medical works, these groups of drugs are associated with special effects on the person consuming them. They are attributed with the strongest effects of all vitalization medicines, including bestowing extreme longevity, and even special powers.[12] But in alchemical literature, their main function is to facilitate the various steps in processing mercury.

According to the *Rasaprakāśasudhākara*, the divine drugs aid in the binding or solidification (*bandhana*) of mercury; the essence drugs are used for the procedures of calcination (*māraṇa*), digestion/amalgamation (*jāraṇa*), and fixation (*niyamana*); the great drugs are applied in solidification and calcination; and the drugs for success in all mercurial operations, though they can also be used in lieu of alchemically processed mercurial elixirs.[13]

The idea of treating one material with another to make it fit for purpose is taken on in ayurvedic medicine from about the eleventh century.[14] There are precursors of the kinds of procedures applied in alchemical operations in older ayurvedic texts. Early examples include the processing of raw iron to make iron tonics, already described in the earliest ayurvedic treatises, the *Carakasaṃhitā* and the *Suśrutasaṃhitā*.[15] However, the described procedures are much simpler, and, more significantly, they are not explained by the medical writers as being necessary for getting rid of contaminants or unwanted characteristics of materials. Once ayurvedic works integrated more recipes for medicines based on inorganic substances, and mercury in particular, they also adopted the alchemical concept of processing materials before use. First descriptions of procedures were integrated directly into

[11] The lists of divine and great herbs and so on are not even the same in the medical works, and there are also variations of the lists in alchemical works. Meulenbeld 1999–2002 records lists of divine and other herbs in both medical and alchemical literature, noting differences.

[12] See, for example, the translation by Wujastyk 2003: 125–131 of a passage on the use of the divine drug of Soma in the *Suśrutasaṃhitā*. Here, the intake of the drug is attributed with letting the consumer attain a lifespan of ten thousand years, an indestructible body, extreme strength, and the ability to go into any realm at will.

[13] The drugs and their properties are described in the ninth chapter of the *Rasaprakāśasudhākara*.

[14] The ca. ninth-century medical treatise *Kalyāṇakāraka* by Ugrāditya may be the earliest medical text to describe alchemical procedures for mercurials. There are, however, some reasons to suspect that the chapter dealing with mercurials, which describes advanced procedures and the use of alchemical apparatuses, may be a later addition to the text. On the adoption of alchemical procedures in ayurvedic literature, see Wujastyk 2019.

[15] These works date to the early centuries CE.

recipes for medicines. For example, a recipe in the eleventh-century ayurvedic work Cakradatta calls for the calcination (*māraṇa*) of iron as one of the steps for making its "immortality-essence iron (tonic)" (*amṛtasāralauha*).[16] In some cases, procedures were not described, but seem to have been a requisite for the products used. For example, the eleventh-/twelfth-century medical treatise *Vaṅgasenasaṃhitā* mentions purified sulfur (*śuddhagandhaka*) in a vitalization tonic recipe, without giving instructions on how to purify it.[17] In later works, procedures for making substances fit for purpose are given more attention and form separate chapters in some medical works.[18] However, even these medical works do not outline a formal list or sequence of procedures and also do not use the terms of the *rasakarman*/*saṃskāra*.

Sequences of Alchemical Operations

As noted above, the list of eighteen procedures in the *Rasaprakāśasudhākara* follows that of the *Rasahṛdayatantra* with some minor changes in spelling or expression. This is an area in which alchemical works display quite a lot of variation, differing in both the number and the details for the procedures, adding or subtracting steps, renaming or reordering the procedures (see the Appendix to this chapter for lists of *saṃskāra* in the alchemical works discussed in this volume).

The *Rasaprakāśasudhākara*'s list of procedures creates the initial impression of an alignment with the *Rasahṛdayatantra*'s alchemical program. However, a closer look at the treatise's contents shows this alignment to be only partial. That is, there is much overlap and parallel in the procedures applied to mercury and other substances, but the *Rasaprakāśasudhākara*'s overall aims differ substantially from the older work in regard to the anticipated effects on the human user of elixirs. While the *Rasahṛdayatantra* describes immortality, special powers, and a transcendent, god-like state as the possible outcomes of the intake of elixirs in the *rasāyana* process, the *Rasaprakāśasudhākara* seems to only pay lip service to such aims. It lays

[16] *Cakradatta* 66.34–125.

[17] *Vaṅgasenasaṃhitā rasāyanādhikāra* 117.

[18] The ca. thirteenth-/fourteenth-century *Śārṅgadharasaṃhitā* devotes its eleventh chapter and part of its twelfth chapter of the middle section (*madhyamakhaṇḍa*) to the subject of purifying and readying inorganic and organic substances for medicinal use. The ca. sixteenth-century medical treatise *Bhāvaprakāśa* devotes a full chapter to the purification and calcination of metals and minerals in the second part of its first section (*pūrvakhaṇḍadvitīyabhāga*).

out the effects of eating the mercurial elixir in its final verses (160–165) of chapter 1, and the emphasis is clearly on their body-enhancing and health benefits. The eradication of disease is mentioned three times, the attainment of special powers (*siddhi*) once, being freed of wrongdoing (*pāpa*) and the attainment of the highest condition (*paramagati*) once. The idea of attaining the highest condition (as well as health, longevity, wealth, and righteousness) is reiterated in the last verse of the third chapter, which deals with making mercury ash. However, all other parts of the work that describe the intake of mercurials only mention health benefits (anti-ageing, sexual stamina, virility) and the cure of diseases, using ayurvedic disease categories throughout.[19] The application of elixirs thus does not seem to be directed at spiritual or transcendent aims, but rather at mundane uses.

However, it would be wrong to describe the *Rasaprakāśasudhākara* as a medical work, as it also has a strong emphasis on making alchemical gold, silver, and gems. The text only hints at this in its first chapter. Even though ten verses (147–156) are dedicated to the subject of transmutation (*vedha*), information on what transmutation takes place is sparse. Only verse 155 spells it out: "Metals become gold and silver through contact with the vapor (of mercury). This is decisively known as 'transmutation through mercury vapor.'"[20] By contrast, the eleventh chapter describes making gold, silver, artificial pearls, and coral at length. One of the instructions (verse 77–79) states that this gold is suitable to be sold in the market.[21] Notably, a substantial number of the instructions for making gold do not contain mercury. Further, not all instructions describe true transmutation (*vedha*), but merely an alteration of appearance. For example, a recipe based on mercury produces "excellent pure gold" out of copper, but another, mercury-free recipe with chalcopyrite as the main ingredient produces something "resembling gold."[22] Some recipes are for the improvement of the appearance of gold or silver, specifically their color.

[19] Chapter 8 gives more than one hundred recipes of mercury compounds for the treatment of diseases. Chapter 12 lists five recipes for the promotion of sexual stamina. However, all of these are herbal recipes that do not contain processed mercury. Chapter 13 adds a further six recipes for sexual stamina and semen retention. Again, these do not contain processed mercury. One recipe includes cinnabar, and several contain opium and poisonous substances.

[20] *dhūmasparśena jāyante dhātavo hemaraupyakāḥ / dhūmavedhaḥ sa vijñeyo rasarājasya niścitam //*

[21] Other verses in the same chapter mention the marketability of alchemical products, for example verses 61–64 describe the making of market-worthy silver.

[22] *Rasaprakāśasudhākara*, chapter 11, verse 6: *jāyate pravaraṃ hema śuddhaṃ* . . . and verse 8: . . . *hemanibhaṃ bhavet.*

The ideas of both selling alchemical gold and using other materials for imitating precious metals and gems stand in contrast with the *Rasahṛdayatantra*'s exposition of the subject. It presents the ability of alchemically enhanced mercury to produce gold or silver not as an aim in itself, but rather as proof of mercury's readiness and suitability for transforming the human body. Alchemically produced gold is thus merely a side product, and not the main concern of the alchemist. However, other alchemical works give more weight to the making of precious metals and gems, providing similar recipes to the *Rasaprakāśasudhākara* and treating the production or imitation of gold, silver, and gems and the use of mercurial elixirs in elixir regimen (*rasāyana*) or in medical treatment as separate, unconnected issues.[23]

Note on the Text and Translation

The translation and all references to readings from the text are based on the 1923 edition of Vaidya Jādavjī Tricumjī Āchārya, his second edition of the text. This edition can be accessed on archive.org at https://archive.org/details/rasaprakasasudhakara1923.

A digital version of the text is available on Oliver Hellwig's *Digital Corpus of Sanskrit*. However, the electronic text is based on the 1940 edition by Jīvrām Kālidās Śāstri, which significantly differs from Tricumjī's edition in the order of verses, and also, though less substantially, in the wording. Śāstri's edition gives a different order for the eighteen alchemical operations, placing coloring (*raṃjana*) after transmutation (*vedha*). The edition and translation by Joshi (2010) follows Śāstri's edition.

[23] The ca. thirteenth-century *Rasaratnākara*'s *Vādikhaṇḍa* is an early example of this separation of the making of precious materials from the use of mercurial elixirs in medicine or *rasāyana*, dealt with in the *Rasendrakhaṇḍa* and *Rasāyanakhaṇḍa*, respectively. See Meulenbeld 1999–2002: IIA, 653–665 on the contents of the different parts of the *Rasaratnākara*, and IIA, 662–663 on the *Vādikhaṇḍa*'s recipes for gold-making and so on.

Rasaprakāśasudhākara Chapter 1

1. I first bow to Harihara, who, bright with the crescent moon, is handsome, who preserves what is auspicious and is therefore revered by Vishnu, and who by nature generates, sustains, and destroys.
2. And I bow to and honor Sharada, who wears white water lilies, jasmine, and white clothes and is well adorned with a necklace of flawless pearls, and who is worshiped by all perfected beings
3. I at once bow to the elephant-faced Ganapati, who has the luster of a leaf of mica and who has three eyes, who is small of stature with four arms, and whose crown shines brightly with gold and jewels.
4. (I bow to) the highest, who always dwells on the mountain, the great and wonderful one, who eradicates all diseases, who is celebrated by the community of all perfected beings, who removes wrongdoings, who leads beyond worldly existence.

5–6. First, I shall now precisely describe the origin of mercury, and then its refinement through the eighteen alchemical procedures, the fourfold art of fixing mercury and (making it into) an ash, the refinement of metals (*dhātu*),[24] their calcination, and their properties.

7. I will then properly explain the eight great substances (*mahārasa*) and the properties of the subsidiary substances (*uparasa*), their refinement and calcination.
8. I shall also relate in detail the liquefaction and distillation of all of these, the good properties and flaws of gems, and their refinement and calcination.

9–12. I will describe in detail the advancing of penetrative power, coloring, liquefaction, and admixture for mercury and one hundred mercurial formulations; the four categories of herbs known as divine herbs (*divyauṣadhi*), essence herbs (*rasauṣadhi*), herbs for success (*siddhauṣadhi*), and finally, the great herbs (*mahauṣadhi*) as taught; and the apparatuses (*yantra*) and heating devices (*puṭa*), the crucibles, and, concisely, the counterfeiting of metals; the formulations for virility treatment and those for strengthening semen. These are the divisions in this book, neither too concise, nor too detailed.

[24] The metals counted as *dhātu* in the *Rasaprakāśasudhākara* (chapter 4) include gold, silver, copper, iron, lead, tin, and different kinds of alloys such as bell metal and brass.

13. In the western part of the Himalayas, there is a scenic mountain called "Prince Among Mountains" (*girīndra*). Close by, the Lord of Essences (*rasendra*) dwells in a perfectly round pit.
14. A beautiful and well-adorned girl, riding on an excellent horse, came there, looked into the well, and retreated in great haste. The most excellent quicksilver immediately followed and descended in the four directions onto the earth.
15. All around the pit, a round area spread out over a full ninety-six miles[25] was strewn with mercury.
16. It is said that the shining mercury that is extracted from that soil in a distillation device instantly removes disease on its own.
17. The mercury that has sprung from the well and has gone into the four cardinal directions indeed arises of the four (classes) as a twice-born, as one of the ruling class, of the merchant and agricultural class and the serving class.
18. It is the color white, black, and yellow, or indeed red in the eastern, southern, western, and northern region, respectively.
19. White (mercury) should be used for making silver, and the black enhances the body, the yellow-colored produces gold, and the red eradicates disease.
20. When all (kinds of mercury) are mixed together they are completely efficacious, and consumed, they will eradicate all diseases and bestow all special powers.
21. Having thus understood the origin of mercury,[26] one is indeed not afflicted by diseases. It is suitable after it has been cleansed, calcined, properly potentiated, and its penetrative power has been advanced.

22–25.1. One should perform steaming as well as trituration, thickening and raising, condensing, and reviving it, properly fixing it, and stimulating it well, and then measuring the morsel of mica and feeding it (to the mercury), and indeed in due course, enclosed liquefaction, external liquefaction, and the mentioned digestion, coloring, setting in motion/potentiation, advancing the penetrative power, and the operation of transmutation, and

[25] = twelve *yojana*.
[26] Or: "what is produced from mercury."

the partaking of mercury: these are the eighteen alchemical operations. Thus, I have here given their names.

25.2–26. The best of physicians have taught that mercury has five flaws: dirt, poison and fire, disorientation, and excitation, respectively.[27] They cause stupor, death, disorientation, boils, and giddiness.

27. There are seven contaminants of mercury. I will relate the facts about them. I will give their names according to the *Teachings of the Goddess* (*Devīśāstra*).[28]

28. There are seven contaminants in mercury, which arise from earth, rock, water, copper, iron, lead, and tin, and these are furthermore highly poisonous.[29]

29–30.1. If these twelve flaws are not expelled by the twice-born, they will not attain perfection through mercury. Instead, the (flaws) will (act) like the god of death. For that reason, the best of physicians should eradicate the faults.

30.2–31.1. Therefore, one should carry out a steaming procedure and the steaming of mercury is properly carried out by alchemists on an auspicious day, using a swing-apparatus (*dolāyantra*).

31.2–32. The two alkali together with a sour substance and also the five salts, the three pungent substances (i.e., black pepper, long pepper, and dried ginger), and also the three myrobalans mixed with leadwort and yellow sulfate of iron and alum.[30]

33. One should rub all these together, equal amounts of all these drugs to an eighth part of mercury. Having prepared a crucible out of them, one should place the mercury into the center.

34–35. One should cover this with three layers of cloth and birch bark. The mercury bundle is fixed to a piece of wood with string and suspended in a vessel filled with sour gruel. One should

[27] The *Rasaprakāśasudhākara* shares this list of five faults with the *Rasendracūḍāmaṇi* (15.23) and the *Rasādhyāya* (17–18), see Hellwig 2009: 119–120.

[28] The *Devīśāstra* is referred to in several alchemical works, including the *Rasaratnasamuccaya* and the *Rasayogasāra*, and in various commentaries on alchemical works. P. C. Ray suggested that the *Devīśāstra* is the same as the *Rasārṇava*. See Meulenbeld 1999–2002: IIA, 744 and IIB, 733n170.

[29] *Rasaprakāśasudhākara*'s inclusion of copper and iron in its list of seven contaminants, or "coatings" (*kañcuka*), follows the list first given in the *Rasendracūḍāmaṇi* 15, 24. Other works do not include copper and iron, but propose that lead and tin form two *kañcuka* each. See Hellwig 2009: 118–120.

[30] I have translated *puṣpakāsīsa* as "yellow sulfate of iron," and *saurāṣṭrā* as "alum," following Hellwig 2009: 195 and 353, respectively.

properly steam it for three days. This explains the sweating procedure.

36. Now I will explain the procedure of trituration (*mardana*) through which mercury becomes most pure in detail and with precision.

37–39. One should triturate mercury in a mortar (*khalva*) for three days.
Now I shall relate the dimensions of the mortar for this.
Namely, sixteen fingerbreadths length and nine fingerbreadths height. One should make it six fingerbreadths wide and deep, and two fingerbreadths wide at the neck. Its internal surface should be made utterly smooth. The pestle should be crescent-shaped and ten fingerbreadths long.

40–41. 240 grams[31] of mercury should be purified by the best of physicians in this. One should triturate it with the herbs prescribed for steaming and with sour gruel for three days. In order to eradicate external contaminants once and for all, one should subsequently wash the king of essences with hot gruel.

42. In the following, I will explain the procedure for thickening mercury.
The thickening procedure expels flaws and removes the seven external contaminants.

43–45ab. One should triturate natron, barley-straw alkaline salt, and also the five kinds of salt and all the sour drugs[32] with mercury for three days in a mortar until it dissolves into a paste (*naṣṭatvam*).[33] Because of the removal of its intrinsic characteristics, this is called thickening here. It becomes flawless and emerges as one that severs knots (*granthibheda*).

45cd–46. Now the best of physicians should diligently undertake the procedure of raising mercury (*utthāpana*) according to the mercurial discipline (*rasaśāstra*). It should be steamed in a swing-apparatus for three days as before.

[31] = five *pala*.

[32] This refers to a class (*varga*) of plants with acidic leaves or fruits, which includes lime, orange, pomegranate, tamarind, sorrel, and others.

[33] According to Hellwig 2009: 261–262, *naṣṭatvam* stands for *naṣṭapiṣṭi* here, describing a state of mercury in which it has become a paste and in which its natural characteristics such as its color and volatility have disappeared.

47. After it has been triturated in the heat of the sun for a day on a rock. Raising (mercury) will properly remove flaws caused by the thickening.

48–49. I will describe the great procedure of distillation (*pātana*) in full detail. It is said that distillation, which removes the flaws of mercury, has three parts: upward distillation, downward distillation, and transverse distillation.

50–51. The characteristics of an upward distillation device are now described here.

An earthen bowl should be made that is six fingerbreadths high, with a diameter of seven fingerbreadths at its opening and thirty fingerbreadths all around. A second suitable bowl of that size should be made.

52. One should triturate mercury with the two alkali and asafoetida as well as the five salts mixed with the group of sour drugs.

53. One should smear the [inside of the] suitable lower bowl with this paste, and, having placed the top one upside down on it, one should join them together into a hemispherical bowl.

54. One should seal the joint with ash and salt. Having placed the bowl on a fireplace, one should then make a fire of grains there.

55. On top (of the spherical bowl), place a water container for up to twelve hours.[34] Having ascertained that (the bowl) has cooled down by itself, one should gather the mercury that has risen up.

56ab. Thus, the upward distillation device has been described.

56cd–58ab. The bowl described above is placed upside down in a hole in clay soil. One should light a fire above for nine hours.[35] The mercury will fall down. Thus, the downward distillation device is described by an expert on mercury.

58cd–60ab. One should triturate the king of essences (i.e., mercury) with the drugs mentioned earlier. One should place the mercury in a pot that lies horizontally. In its opening, one should place another, smaller pot with a hole in its middle (*udara*), and in that hole a metal pipe.

[34] = four *yāma*, one *yāma* being a period of about three hours.
[35] = three *yāma*.

60–61ab. The best of physicians should place the pipe into a water bowl. A wise person should light a strong fire underneath the apparatus with the mercury. After nine hours, the mercury will have fallen transversely.

61cd–62ab. I have thus properly described the three kinds of distillation device that are easily made according to the teaching of the doctrine of mercury.

62cd. And now I shall describe the procedure of reviving mercury.

63–64. One should place 480 grams[36] of salt and 2.7 litres[37] of water and the mercury, which has been freed of its flaws into a pot that is covered with a lid and properly sealed with clay.

65. One should place this in a windless and isolated spot for three days. Carry out the reviving in exactly this way, o king of physicians.

66–67ab. Next, I shall describe the fixation of mercury through which the volatility of the king of essences ceases. The best of mercury is placed in a pot together with water and salt. Once it has been steamed for three days, it will emerge full of vigor.

67cd–68ab. And now I will speak of the procedure of stimulation (*dīpana*) of the king of essences, through which its hunger and pervading power properly arise.[38]

68cd–69. Mustard (seed) with sea salt, black pepper, moringa (*śigru*) and borax, iron sulfate, and alum together with sour gruel: having steamed [mercury] together with these for three days, one should then triturate it with alkali.

70–71ab. In this way, stimulation is quickly achieved. If this is done properly, intensity, reactivity, pervading power, hunger, and particularly strength arise.

71cd–72ab. I shall fully explain the procedure of producing a mouth, which is actually a type of stimulation (*dīpana*) and a proper act of mastering mercury.

[36] = ten *pala*.

[37] = 3 *prastha*. There are different definitions of a *prastha*, varying between 32 *pala* and 6 *pala*. The *Rasaprakāśasudhākara* does not define *prastha*. I here follow Hellwig 2009: 25–26, who uses the definition in the *Rasaratnasamuccaya*, according to which 1 *prastha* = 2 *māṇikā*; 1 *māṇikā* = 2 *kuḍava*; 1 *kuḍava* = 2 *prasṛta*; and 1 *prasṛta* = 2 *pala*. Therefore, 1 *prastha* equals 16 *pala*. As a liquid measure, one *pala* is a bit over 2 imperial ounces = ca. 57 millilitres. A *prastha* would therefore be ca. 912 millilitres.

[38] "Hunger" (*bubhukṣā*) refers to mercury's ability to assimilate other substances.

72cd–73ab. Mercury should be rubbed together with an eighteenth part of gold and lime juice in succession, one after the other.

73cd–74ab. After it has been steamed with alkali, salts, sour substances, and sour gruel, and washed with sour gruel, it develops a mouth to devour.[39]

74cd–75ab. It should then be triturated in a heated iron mortar on a stove with two parts of gold and a sixteenth part of essence of chalcopyrite.

75cd–76. After it has become amalgamated, one should triturate it with vinegar, chick pea alkali, and lemon and citron water for twenty-one days, each for three hours.[40]

77–78. There is no doubt that it will properly develop a mouth like that of a demon.[41] I have thus related the second stimulation procedure. This describes the eight mercurial procedures in the context of procedures for sustaining the body. And now the ten procedures acting on the body and on metals.

79. As the apportioning cannot be done without measuring, I shall explain the above-mentioned procedure of measuring.[42]

80. A sixty-fourth part of a seed (*bīja*) to the mercury gives it a mouth.[43]

Afterward, one should add a thirty-second part of the best of seeds.

81. Then, one should add a sixteenth part of the seed. The best of physicians should add an eighth part (of the seed) to the mercury.

82. Then a fourth part of the morsel should be added. And next, one should apply an equal amount of the morsel.

83. Together with a sixteenth part of the special salt (*viḍa*), the mercury becomes hungry. Once the morsel has been digested, (the mercury) is distilled together with the special salt.

[39] i.e., it develops an ability to assimilate metals and minerals.

[40] = one *yāma*.

[41] A "mouth like that of a demon" (*yātudhānamukha*) probably refers metaphorically to the ferocious appetite associated with the *yātudhāna* (~ spirits, demons) and therefore to the heightened ability of mercury to assimilate other substances.

[42] The apportioning refers to determining the quantity of the substance that should be added to the mercury. See Hellwig 2009: 159–160 and Dash 2003: 69.

[43] Giving mercury a "mouth" means to enable or enhance its ability to eat, i.e., assimilate other substances. See Hellwig 2009: 251–252.

84. Iron sulfate, rock-salt, sea salt, sochal salt, and alum with an equal amount of sulfur: this special salt will enhance digestion (i.e., assimilation of other substances).

85. I shall now explain the intake of mica, which is properly done in a hemispherical metal bowl according to the "skull-method" (*karoṭividhinā*).[44]

86. One should without doubt feed morsels of mica to mercury joined into a compound with a special salt (*viḍa*) in a water device (*jalayantra*).[45]

87. I shall explain the characteristics of a water device in the following.

A well-rounded iron bowl with 7.68 litres[46] of water in it.

88–89. Inside it, one should place a firm sphere made of two iron bowls, with the purified mercury placed inside the iron sphere, together with a sixteenth part of the special salt (*viḍa*), with care, and a sixty-fourth part of mica essence in the sphere as well.

90. One should then seal the joints of the sphere with an impenetrable water-proof paste (*toyamṛd*) or fire-proof paste (*vahnimṛd*).[47]

91. Having set the device, filled with hot water, onto a stove, one should increase the fire bit by bit over half a day.

92. By doing it this way, one will cause the portions of morsels to be assimilated, not otherwise. Through this method, one will cause it to assimilate the six morsels quickly.

93ab. Once the mica essence has been assimilated, it bestows success in all undertakings.

[44] Hellwig 2009: 160–n308 suggests that this may refer to the resemblance of the shape of each of the two bowls that together form the hemispherical iron bowl to the top of a human skull. Perhaps this would mean an oval, quite shallow bowl shape. Meulenbeld 1999–2002: IIB, 659 translates *karoṭi* as "shallow iron cup."

[45] I have translated *viḍa* as "special salt" here, since there is more than one kind of *viḍa*. On the various preparations that go under the name of *viḍa* and are meant to help mercury digest, i.e., assimilate the substances fed to it, see Hellwig 2009: 446–448. The *Rasaprakāśasudhākara* gives a recipe for a *viḍa* in 1.92, which is presumably the one meant to be used here.

[46] = three *āḍhaka*.

[47] See Hellwig 2009: 167 and 418 for *jalamṛd*, a water-proof paste containing chalk, rust, molasses, and a decoction of acacia; and for *vahnimṛtsnā* (= *vahnimṛd*), a fire-proof paste of chalk, salt, rust, and milk, respectively.

93cd–94ab. Now I shall describe the quality-enhancing feeding procedure of enclosed liquefaction that truly makes the king of essences succeed.

94cd–95. One should melt equal amounts of chalcopyrite essence and mica essence and copper extracted from brass in an acacia (coal) fire, using a pair of bellows until (only) mica remains.

96–97ab. One should let an equal amount of mercury digest that essence of mica. Through this procedure, a threefold digestion takes place in the mercury.

This is the (procedure of) digestion through enclosed liquefaction as described by the best of physicians.

97cd–99. The rules for external liquefaction are explained according to the path of the gurus.

The essence of mica that is melted in a crucible together with the sap of veldt grape (*vajravallī*)[48] and sochal salt takes on the appearance of mercury. And it is thought that through the admixture of liquified mica with equal amounts of mercury, it (i.e., the mercury) becomes fixed. This is called external liquefaction.[49]

100–101ab. The procedure of applying external liquefaction succeeds through devotion to Śiva, through the favor of the guru, and through constant worship of Mahābhairava. And through homage to the two auspicious ones, the external liquefaction will indeed succeed.

101cd–102ab. Now I shall explain the procedure of digestion in detail. One should mix equal amounts of mica and essence of chalcopyrite and melt them together.

102cd–103ab. Once the remainder of mica has been produced, one should let its essence be digested in the mercury. Thus, one should prepare the essence of mica with the two putrid metals (tin and lead).

[48] The identification of *vajravallī* as veldt grape is derived from an entry in the FRLHT's Envis Centre on Medicinal Plants online plant database at http://envis.frlht.org/ (accessed on April 14, 2020). There is no entry for *vajravallī* in Chopra, Nayar, and Chopra 1956.

[49] *Druti* is meant to effect lasting liquefaction of both the added substance and the mercury. However, see Hellwig 2009: 122–123 on the interpretations in various alchemical works and commentaries of the binding effect of *bāhyadruti* on mercury. This binding effect means the mercury is no longer liquid, and the *bāhyadruti* of verse 99 should perhaps no longer be translated as "external liquefaction." However, the mica is dissolved within the mercury.

103cd–104ab. According to the rules of metallurgy (*dhātuvāda*), this acts on metals, not on the body. Tin and lead are terrible (poisons) that should not be taken continuously.

104cd–105. A quarter of a gram[50] of the prepared essence of mica that has been ground and resembles dust is thrown into the mouth of the hungry mercury. A quarter of eight grams[51] of mercury is recommended.

105cd–107. Vinegar placed in a copper pot together with rock-salt and also milk is triturated for up to two days. It is taught by knowledgable people that it will become a black paste similar to collyrium.[52] One should rub mercury with this paste for three days in a heated mortar.

108–109. And after it has been washed with sour gruel, one should put it in a glass bowl, placing the special salt in the amount of an eighth part to the mercury above and below, and forming a sphere (*saṃpuṭa*) around it. One should cover the opening with birch bark and fasten it with cord.

110. One should then cover the sphere with cloth and steam it in a swing (-apparatus) with cow urine, vinegar, and sour gruel every day.

111–112. One should then wash the mercury with hot sour gruel in either a stone, metal, or glass bowl and place it into four layers of firm cloth. Through pressing it down, the mercury passes into another bowl.

113. If all of it is mercury, the mica was digested properly. One should repeat this. Measuring the morsel again, an excellent seed of mica should be given.

114. One should let (mercury) digest all with eight morsels according to the path of the gurus. Having done it in this way, the mica will always be digested quickly in the mercury.

115. I have properly carried out the digestion procedure by my own hands, this is not hearsay. Mercury will become a

[50] = one *valla*. This weight is approximate. One *valla* may be equal to a weight of one and a half, two, or three *guñjā*. I have used one *valla* = two *guñja* here, which is about a quarter of a gram.

[51] = one *gaḍyāṇa*. This weight is uncertain. One *gaḍyāṇa* is understood as thirty-two *guñjā* (= four grams) by some, and sixty-four *guñjā* (= eight grams) by others.

[52] I have translated *tuttha* as collyrium, since the end product is described as a black paste. However, see Hellwig 2009: 406–407 on the difficulties of identifying *tuttha*.

"stick-holder"[53] after it has fully digested an equal amount of mica.

116. And it is called a child that is efficacious in some operations and I also call it that when it has digested twice or three times its amount of mica.

117. When it has digested four times its amount of mica, I call it a youth. When it has digested five times its amount of mica, the best of essences is also a youth.

118. When it has digested six times its amount of mica, the best of essences becomes mature. When it has digested seven or eight times its amount of mica essence, the mercury becomes very aged.

119. It induces all special powers. Through this method, the mercury itself will digest all metals.

120. I shall now speak of the dyeing of mercury.
After it has been dyed and had its penetrative power increased, (mercury) manifests as the god Maheśvara.

121. Iron and pyrites are dyed with zinc ore, and also with the group of red substances, according to the tenets of the discipline.

122. The dyeing of mercury, in turn, is done with sulfur dye, with copper, with red glass, and with red rock-salt.

123–124ab. One should dye mercury in an enclosed crucible with pyrites, and so on, and one should dye it with sulfur dye using a brick apparatus (*iṣṭikāyantra*),[54] and with zinc ore dye through employing a balance apparatus (*tulāyantra*).[55]

124cd. One can always dye mercury through trituration with wrought iron (*tīkṣṇa*) filings.

125. One should keep (mercury) made into a paste with copper for seven days. This brings about the dyeing of mercury without doubt.

[53] The condition of being a "stick-holder" (*daṇḍadhārin*) as a sign that the digestion procedure was successful is explained in the *Rasendracūḍāmaṇi* (16, 33), which describes a kind of test in which a blade of grass is placed on the mercury and stays on top, i.e., does not sink. This makes the mercury a "stick-holder." See Hellwig 2009: 102.

[54] See Hellwig 2009: 165–166 on the brick apparatus (*iṣṭikāyantra*). Mercury is filled into a hollowed-out brick, topped with sulfur and covered and then heated in a pit.

[55] On the balance apparatus (*tulāyantra*), see Hellwig 2009: 403–404.

126. One should make an earthen crucible, smear it with the red substances,[56] and place mercury into it. Through roasting, the dyeing will occur.

127. I have briefly explained the dyeing of mercury. However, many ways of dyeing are taught in the tradition.

128. Now, I shall describe the procedure of augmenting transmutation, the potentiation (*sāraṇa*) of all procedures, which will produce the great special powers.

129–131ab. One should make an earthen crucible similar to a datura flower, eight fingerbreadths deep, with a width of four fingerbreadths at the opening, dried, and made very smooth on the inside, and another crucible as a cover with a deep opening. One should deposit the purified mercury that is ready to digest inside the crucible.

131cd–133. One should combine the fat of fish, turtles, frogs, leeches, sheep, and pigs into potentiating oil (*sāraṇātaila*), as well as honey, earthworm cast, crow droppings, feces of locusts, and so on and the ear wax of buffaloes and make them into a paste with a sixteenth part of mercury.

134. Having passed it through a cloth, one should add it to sesame oil, and one should thoroughly cook this oil made for the purpose of potentiation in this sesame oil.

135. Having prepared a seed and the paste mix, one should place them onto the crucible and close the mouth of the crucible with a second lid.

136–137. One should seal the pair of crucibles with ash and salt. Having buried the three parts of the crucible, one should place soil on top and roast it well with a strong fire with firm charcoal. In this way, the digested seed is able to dissolve inside the mercury.

138. The potentiated mercury becomes fixed and develops good qualities. First, after it has been made to digest and has been potentiated, it produces all special powers.

139–140ab. If it is not made to digest and potentiated, how can it become a transmutative agent? I have witnessed the most excellent potentiation procedure according to the teaching of the

[56] The group of red substances (*raktavarga*) are specific organic substances that, however, vary from treatise to treatise. See Hellwig 2009: 317.

guru and have properly performed it from practical experience, not from hearsay.

140cd–141ab. Now I will address the procedure of advancing the penetrative power of mercury. This is, however, done according to a book, not from experience. One should properly advance the penetrative power of mercury.

141cd–143. One should triturate equal amounts of these substances for a day: the ear wax of buffaloes and breast milk mixed with borax; also aconite and cinnabar, zinc ore, *kamala*,[57] and magnetic iron, red velvet mite and pigeon pea, pyrites and crow feces. One should place mercury into the middle of the paste, (surrounded) above and below by it.

144. One should then add glass powder and melt it inside a covered crucible (*andhamūṣā*). Through this procedure of advancing the penetrative power, mercury penetrates instantly.

145ab. This most excellent procedure of advancing the penetrative power was described by king Nandi.

145cd–146ab. Chalcopyrite essence and purified lead: these seeds always effect the advancing of the penetrative power of mercury without doubt.

146cd–147ab. Now I shall explain the method of transmutation in detail. Through understanding it alone, a man becomes one who knows transmutation.

147cd–149. Datura oil, opium, and oil of the intellect tree, false daisy oil, aconite, and nutmeg oil, oleander root oil, and elephant creeper oil: know that these and other oils produce transmutation. Prepared mercury that is triturated with them becomes an agent of transmutation.

150. The five kinds of transmutation are thought to be transmutation through daubing (*lepavedha*), pouring (*kṣepa*), as well as transmutation through a lance (*kuṃtavedha*), by smoke (*dhūma*) and transmutation through sound (*śabdavedha*).

[57] Many plants go under the name of *kamala*. Chopra, Nayar, and Chopra 1956 has two entries for *kamala* on pages 160 and 174, respectively. The first is Mallotus philippensis Muell. Arg., an evergreen tree, whose roots, stems, and leaves contain a poisonous acid. The second is Nelumbo nucifera Willd., the Indian lotus.

151–152ab. One should roast thin sheets of copper, and of silver, daubed with the (mercury-) paste in a covered crucible. Once this has cooled down, and has been removed, it is called "transmutation through daubing" (*lepavedha*).

152cd–153ab. One should pour the mercury into melted copper or silver. Through this, it is immediately transmuted. This is called "transmutation through pouring" (*kṣepavedha*).

153cd–154. One should melt lead, silver, and copper, and other such substances. One should add mercury that was melted in another vessel. Transmutation is induced. The expert of the discipline said that "transmutation through a lance" (*kuntavedha*) will occur.

155. Metals become gold and silver through contact with the vapor (of mercury). This is decisively known as "transmutation through mercury vapor."

156. Through physical contact with the best fixed mercury, a sound is produced and the transmutation that is effected through this is called "transmutation through sound."

157. I shall now describe the eighteenth procedure, the intake of mercury, carefully and in detail.

158. Mercury that is taken with care according to the tradition will bestow special powers. If it is partaken of in another way, it will kill a man like poison.

159. First, after emesis has been induced, one should undertake purgation. Then, one should eat calcined mica and afterward, the intake of mercury takes place.

160. The best of mercury that has been properly purified always optimizes the body and metals. When it is partaken of, it will eliminate all diseases and bestow all the special powers.

161. An amount of mercury that can transform eight grams of metals will, when eaten, eradicate diseases of the body.

162–163. Experts in alchemical procedures mention these amounts: mercury is measured in the amount of a mustard seed, a beautyberry seed, a field mustard (seed), a mung bean, a black gram, a rosary pea, a chickpea, and a quarter of a gram.[58] It should be eaten together with some betel leaf as a vehicle.

[58] = one *valla*.

164. When mercury is partaken of in this way, it is said to remove all diseases. One will be freed of all wrongdoings and attain the highest condition.

Bibliography

Ācārya, Yādavaśarma Trivikrama. 1988. *Vaidyavaraśrīcūḍāmaṇisaṃgṛhītā Rasakāmadhenuḥ (upakaraṇapāda-dhātusaṃhrahapāda-sūtakriyāpādeti Pādatrayātmakaḥ Prathamo Bhāgaḥ) [the Rasakāmadhenu of Śrīcūḍāmaṇi. Part 1: Pādas 1–3]*. Vārāṇasī and Dillī: Caukhambhā Oriyanṭāliyā.

Āchārya, Jādavjī Tricumjī. 1923. *śrīyaśodharaviracito rasaprakāśasudhākaraḥ. Ācāryopāhvena trivikramātmajena yādavaśarmaṇā saṃśodhitaḥ prakāśitaśca. Rasaprakāsha Sudhākara by Yashodar. Edited and published by Vaidya Jādavji Tricumjī Āchārya*. 2nd edition. Bombay: Āyurvedīya-Granthamāla.

Bhāvaprakāśa. See Srikantha Murthy 1998–2000.

Cakradatta. See Sharma 2002.

Chopra, R. N., S. L. Nayar, and I .C. Chopra. 1956. *Glossary of Indian Medicinal Plants*. Delhi: Council of Scientific & Industrial Research.

Dash, Bhagwan. 2003. *Alchemy and Metallic Medicines in Ayurveda*. New Delhi: Concept Publishing Company.

Hellwig, Oliver. 1999–. *DCS: Digital Corpus of Sanskrit*. http://www.sanskrit-linguistics.org/dcs/.

Hellwig, Oliver. 2009. *Wörterbuch der mittelalterlichen indischen Alchemie*. Groningen: Barkhuis & University of Groningen, University Library.

Hellwig, Oliver. 2014. "Materialgruppen in der indischen Alchemie." *Zeitschrift der Deutschen Morgenländischen Gesellschaft* 164 (2): 451–468.

Joshi, Damodar. 2010. *Rasaprakāśa Sudhākara. Sanskrit Text with English Translation and Appendices*. New Delhi: Indian National Science Academy.

Meulenbeld, Gerrit Jan. 1999–2002. *A History of Indian Medical Literature*. 5 vols. Groningen: E. Forsten.

Meulenbeld, Gerrit Jan. 2000. "Basic concepts of Indian Pharmacology." In *Studies on Indian Medical History*, edited by G. Jan Meulenbeld and Dominik Wujastyk. Delhi: Motilal Banarsidass.

Michaels, Axel. 2017. "Rites of Passage: *saṃskāra*." In *The Oxford History of Hinduism: Hindu Law: A New History of Dharmaśāstra*, edited by Patrick Olivelle and Donald R. Davis, 86–98. Oxford: Oxford University Press.

Rasahṛdayatantra. See Rasaśāstri 1989.

Rasakāmadhenu. See Ācārya 1988

Rasaprakāśasudhākara. See Joshi 2010; Āchārya 1923

Rasaśāstri, Daulatarāma. 1989. *Rasahṛdayatantram. Śrīmadgovindabhagavatpādaviracitaṃ. Caturbhujamiśraracita "Mugdhāvabodhinī" Saṃskṛta ṭīkā tathā Hindī vyākhyā sahitam, vyakhyākṛtṛ Daulatarāma Rasaśāstrī*. Vārāṇasī: Caukhambhā Oriyanṭāliyā.

Śārṅgadharasaṃhitā. See Srikantha Murthy 2006.

Sarma, R. S., and R. Sahai. 1995. "Gushing Mercury, Fleeing Maiden: A Rasaśāstra Motif in Mughal Painting." *Journal of the European Āyurvedic Society* 4: 149–162.

Saxena, Nirmal. 2004. *Vaṅgasena Saṃhitā or Cikitsāsāra Saṃgraha of Vaṅgasena. Text with English Translation, Notes, Historical Introduction, Comments, Index and Appendices by Nirmal Saxena*. 2 vols. Varanasi: Chowkhamba Sanskrit Series Office.

Sharma, Priya Vrat. 2002. *Cakradatta (Sanskrit Text with English Translation). A Treatise on Principles and Practices of Ayurvedic Medicine*. Varanasi: Chaukhamba Publishers.

Srikantha Murthy, K. R. 1998–2000. *Bhāvaprakāśa of Bhāvamiśra, Text, English Translation, Notes, Appendeces [sic] and Index, Translated by K.R. Srikantha Murthy.* 2 vols. Varanasi: Krishnadas Academy.

Srikantha Murthy, K. R. 2006. *Śārṅgadhar-Saṃhitā (A Treatise on Āyurveda) by Śārṅgadhara. Translated into English by Āyurveda-Vidwān Prof. K. R. Srikantha Murthy.* Varanasi and Delhi: Chaukhamba Orientalia.

Vaṅgasenasaṃhitā. See Saxena 2004.

White, David Gordon. 1996, *The Alchemical Body, Siddha Traditions in Medieval India.* Chicago and London: University of Chicago Press.

White, David Gordon. 2017. "Variations on the Indo-European 'Fire and Water' Mytheme in Three Alchemical Accounts." *Journal of the American Oriental Society* 137 (4, October–December): 679–698.

Wujastyk, Dagmar. 2019. "Iron Tonics: Tracing the Development from Classical to Iatrochemical Formulations in Ayurveda." *HIMALAYA* 39 (1). https://digitalcommons.macalester.edu/himalaya/vol39/iss1/11.

Wujastyk, Dominik. 2003. *The Roots of Ayurveda: Selections from Sanskrit Medical Writings.* 3rd edition. Penguin Classics. London, New York: Penguin Group.

Appendix: Lists of saṃskāra/rasakarman[59]

Rasahṛdayatantra, chapter 2, 1–2

1) svedana, 2) mardana, 3) mūrchā, 4) uthāpana, 5) pātana, 6) nirodha, 7) niyama, 8) dīpana, 9) gaganagrāsapramāṇa, 10) cāraṇa, 11) garbhdruti, 12) bāhyadruti, 13) jāraṇa, 14) rasarāga, 15) sāraṇa, 16) krāmaṇa, 17) vedha, 18) bhakṣaṇa

Rasārṇava, chapter 10, 10

1) svedana, 2) mardana, 3) cāraṇa, 4) jāraṇa, 5) drāvaṇa, 6) rañjana, 7) sāraṇa, 8) krāmaṇa

Rasendracūḍāmaṇi, chapter 15, 28

1) svedana, 2) mardana, 3) mūrchā, 4) utthāpana, 5) pāta, 6) rodha, 7) niyama, 8) proddīpana, 9) grāsajamāna, 10) cāraṇa, 11) garbhajanitadruti, 12) bāhyajanitadruti, 13) jāraṇā, 14) rāga, 15) sāraṇa, 16) parikrama, 17) vedha, 18) sevana

Rasaratnākara Vādikhaṇḍa chapter 11, 2-3

1) svedana, 2) mardana, 3) mūrchā, 4) utthāpana, 5) pātana (3 kinds), 6) nirodhana, 7) niyāma, 8) dīpana, 9) ānuvāsana, 10) jāraṇa, 11) cāraṇa, 12) garbhadruti,

[59] This show the lists of the procedures of the works introduced in this volume. Other alchemical works also feature lists of *saṃskāra/rasakarman.* For the purpose of this list, I have used the editions in the Digital Corpus of Sanskrit, except in the case of the *Rasahṛdayatantra, Rasakāmadhenu,* and the *Rasaprakāśasudhākara,* whose editions are listed in the bibliography.

13) bāhyadruti, 14) rañjana, 15) sāraṇā, 16) anusāraṇā, 17) pratisāraṇā, 18) krāmaṇa, 19) loha(vedha?), 20) deha(vedha?)

Rasamañjari

No list of saṃskāra

Rasaratnasamuccaya, chapter 11, 15–16

1) svedana, 2) mardana, 3) mūrchana, 4) utthāpana, 5) pātana, 6) rodhana, 7) niyāmana, 8) saṃdīpana, 9-10) gaganabhakṣaṇamāna, 11) saṃcāraṇa, 12) garbhagatadruti, 13) bāhyadruti, 14) jāraṇa, 15) sāraṇa, 16) saṃkrāmaṇa, 17) vedha, 18) śarīrayoga

Rasaprakāśasudhākara, chapter 1, 22–25

1) svedana, 2) mardana, 3) mūrchana, 4) utthita, 5) pātana, 6) nirodhana, 7) niyamana, 8) dīpana, 9) abhrakagrāsamāṇa, 10) cāraṇa, 11) garbhdruti, 12) bāhyadruti, 13) jāraṇa, 14) raṃjana, 15) sāraṇa, 16) krāmaṇa, 17) vedha, 18) sevana

Rasakāmadhenu, section 3 (sūtakriyāpāda), chapter 2, 1

1) svedana, 2) mardana, 3) cāraṇa, 4) jāraṇa, 5) drāvaṇa, 6)ranjana, 7) sāraṇa, 8) krāmaṇa (dehalohavedha mentioned in 15–16)

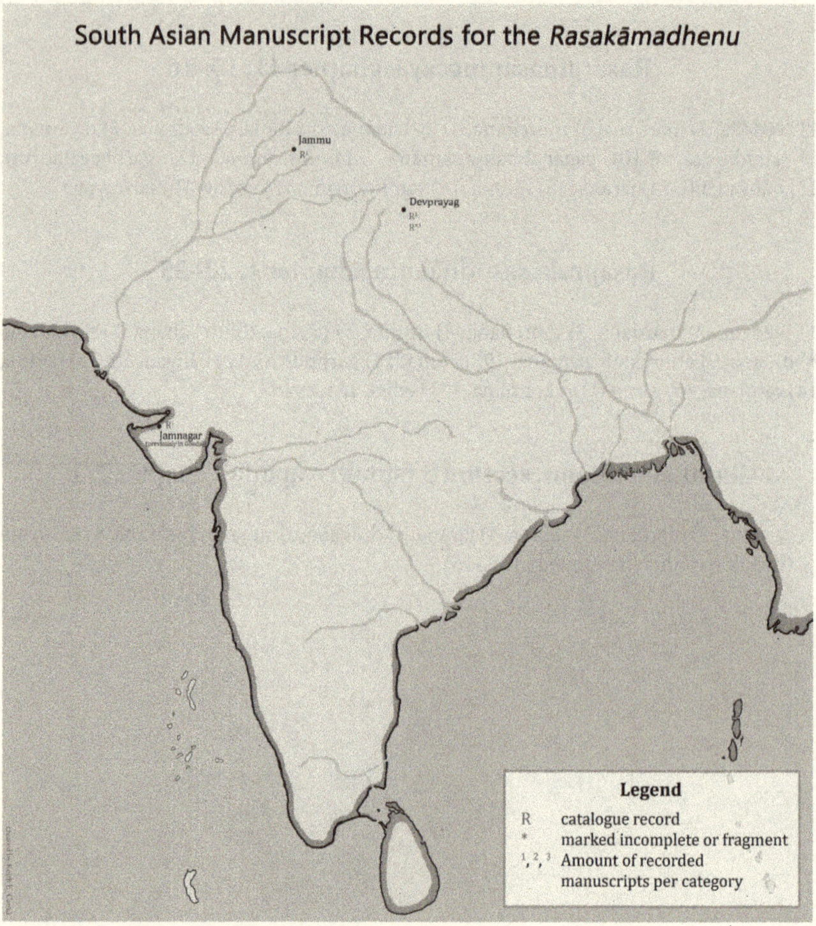

Map 9.1 South Asian manuscript records for the *Rasakāmadhenu*.

9

Alchemical Instruments

Section 1, Chapter 1 of the *Rasakāmadhenu*

Patricia Sauthoff

In the second half of the seventeenth century, Cūḍāmaṇi compiled a systematic treatise on alchemy that cites dozens of earlier texts. Among the most well known of the cited alchemical works are the *Ayurvedaprakāśa*, *Rasahṛdayatantra*, *Rasamañjarī*, *Rasarahasya*, *Rasaratnākara*, *Rasaratnasamuccaya*, *Rasārṇava*, *Rasasāra*, *Rasendracintāmaṇi*, and *Rasendracūḍāmaṇi*. Additionally, some of the quotations also appear in works such as the *Ānandakanda*, *Bhāvaprakāśa*, and *Rasataraṅgiṇī*.[1] Through these citations, Cūḍāmaṇi offered a detailed look at the history and variations on alchemical apparatuses and procedures related to the purification and use of metals, gems, mercury, and other substances. These variations show the creative and experimental nature of the alchemical sciences.

The *Rasakāmadhenu* consists of four sections focused on (1) the instruments of alchemy (*upakaraṇapāda*); (2) the purification and killing of metals, gems, minerals, and salts (*dhātusaṃgrahapāda*); (3) mercurial operations (*sūtakriyāpāda*); and (4) medicine (*cikitsāpāda*).

The text begins with an evocation of great yogi Nārāyaṇa and his beloved Tripurā, also called Tripurasundarī. Most associated with the Śrīvidyā Śakta tradition, Tripurā is connected to Rasāṅkuśī,[2] the goddess of mercury, who is mentioned numerous times in the *Rasaratnasamuccaya*, *Rasaratnākara*, *Rasendracūḍāmaṇi*, *Rasārṇava*, and *Ānandakanda*. As Tripurā she is mentioned only in the *Rasārṇava* (3.25), where she is among a group of goddesses meant to be worshiped before the undertaking of alchemical operations. Though here the beloved of Nārāyaṇa, in the alchemical tradition Tripurā is usually associated with Rasabhairava, a form of Śiva worshiped in

[1] Meulenbeld 2000a: 632–633.
[2] White 1996: 178, 443n60.

Patricia Sauthoff, *Alchemical Instruments* In: *Indian Alchemy*. Edited by: Dagmar Wujastyk, Oxford University Press.
 DOI: 10.1093/9780197805831.003.0010

Kashmir by the Svacchandabhairava cult.[3] Though we learn nothing more about Tripurā in the *Rasakāmadhenu*, it is important to note that originally it was Kāmadeva and not Śiva to whom Tripurī was attached.[4] Though she became associated with Śiva by at least the eleventh century, elements of her link to Kāmadeva are apparent in her iconography.[5] This becomes important for her connection to the *Rasakāmadhenu* as Tripurā's fertile and regenerative powers are key to the completion of alchemical operations. It is also important to note that Tripurā's role as goddess in the *Rasārṇava* may overlap[6] with the end of a period of tantric innovation in Kashmir that reached its height in the tenth to eleventh century.

After invoking the deities, the first chapter of the *Rasakāmadhenu* turns to the instruments and apparatuses necessary for alchemical processes. Cūḍāmaṇi often gives multiple quotations from older alchemical works for the devices, demonstrating non-uniformity in size and shape across various texts. For example, he gives the dimensions of mortars (*khalva*) as:

	Verse 7	Verse 8
Height	16 cm	18 cm
Length	28 cm	28 cm
Width	16 cm	18 cm
Depth	10.5 cm	12 cm
Edge	3.5 cm	3.5 cm
Pestle	8 cm high, 18 cm base	21 cm

In addition to dimensions, Cūḍāmaṇi also lists the materials from which each object is to be made and describes the device's function. He catalogues two types of mortars: *khalva* and *taptakhalva*; twenty-eight types of appliances (*yantra*): *dolā, kinnara, ḍamaru, vidhyādhara, pātana, adhaḥpāta, kacchapa, dīpikā, śrīrañjaka, karuṇākara, siddhasāra, garbhasāra, paramānandamūrtija, pātāla, ākāśa, vāruṇika, jyotir, sādhya, ghaṭa, āpyānaka, saikata, haṃsapāka, vālukā, iṣṭikā, sthūlakūrma, jalakūrma, jalakūrmaprakāra*, and *somānala*; six types of crucibles (*mūṣā*): *vajra, pakva, bhasma, mañjūsa, gola, vajramūsāmṛd*; and two types of seals (*mudrā*): *haṭha* and *madana*. This list is by no means exhaustive,

[3] Ibid.
[4] Golovkova 2020.
[5] Ibid.
[6] See White, this volume, on the second chapter of the *Rasārṇava*.

however it does provide a good overview of the appliances necessary for a variety of alchemical operations. Especially important are the appliances for the calcination of mercury (*ḍamaruyantra*), extraction of cinnabar (*vidhyādharayantra*), distillation (*pātanayantra* and *adhaḥpātayantra*), the production of medicine (*siddhasārayantra*), digestion of sulfur (*kapotapuṭayantra* and *iṣṭikāyantra*), and digestion of mercury and other minerals (*sthūlakūrmayantra*). Similarly, crucibles are used for the purification of silver (*bhasmamūṣā*), distillation (*golamūṣā*), and specific steps in the roasting process. Finally, the chapter ends with details on how to wrap bottles and crucibles with cloth fragments covered in mud and seals to secure the seams between two appliances and create air-tight enclosures.

The remaining chapters in the first section of the text then offer particulars about various fire places, fuels, vessels, plants, and substances to be used, and outline the characteristics of the teacher, initiate, female companion, initiation, workshop of the *rasaliṅga*, and behavioral rules.[7]

Notes on the Translation and Measurements

For this translation I have relied mainly on Ācārya's 1925 edition. However, Mishra and Sharma's 1992 and Śāstri's 1992 editions have proven useful for both line drawings and Hindi commentary. As Meulenbeld notes, these and Śukla's edition contain interpolations added to the first chapter. The Digital Corpus of Sanskrit (DCS) follows Mishra and Sharma. For the citation of parallel verses I have relied upon the DCS and used its numbering. Occasionally, Cūḍāmaṇi notes that a line comes from a text unavailable on the DCS or GRETIL, such as the *Rasasāra*. Unfortunately, I was unable to consult physical editions of these texts due to limitations of time and the shut-down of the interlibrary loan system due to COVID-19. Additionally, I have consulted Hellwig 2009 for cross-references of citations as well as diagrams of the various *yantras*.

I have calculated (and where necessary rounded) measurements using the following conversions:

1 *aṅgula* ≈ 1.763 cm
1 *ṭaṅka* ≈ 4 g

[7] Meulenbeld 2000b: 629.

1 *pala* ≈ 48 g
1 *prastha* ≈ 81 g

I have strayed from the Sanskrit in translating the third-person optative (one should) as a second-person imperative. This is done solely to ease reading for the non-specialist and to present the text in the more familiar language of a manual or recipe. Additionally, Cūḍāmaṇi adds his own prose throughout his text. On occasion his commentary rejects what is said in the quoted passage. However, in chapter 1 the commentary adds very little save for the name of a text or a phrase such as "another apparatus—." Translation of this is often redundant and slows reading and therefore has been omitted here. Textual parallels can be found in the footnotes.

Abbreviations

ĀK	*Ānandakanda*
MuA	*Mugdhāvabodhinī*
Rarṇ	*Rasārṇava*
Rcint	*Rasendracintāmaṇi*
RcūM	*Rasendracūḍāmaṇi*
Rmañj	*Rasamañjarī*
RPSudh	*Rasaprakāśasudhākara*
RRĀ Vkh	*Rasaratnākara Vādakhaṇḍa*
RRS	*Rasaratnasamuccaya*
RRSTīkā	*Rasaratnasamuccayaṭīkā*

Rasakāmadhenu, Section 1, Chapter 1

1. Beloved of Tripurā, arisen from the highest clear light, overcoming disease, made of innate bliss, is Lord Guru Nārāyaṇa himself.
2. This *Rasakāmadhenu*, abridged by Cūḍāmaṇi, conquers with instruments, groups of minerals and elements, mercurial operations, and medicine, which are the four parts.

3–6. Having gathered together the necessary materials, perform the alchemical processes. Mercury, minerals, and metals, mortar, and stone pestle, earthenware vessels and apparatuses, a tube, and an iron instrument. A crucible, blowpipe, cow dung, salts of iron (*sāra*), kindling, a pair of tongs, flat tongs, an earthen vessel, and an iron vessel. Weights and counterweights, cutting instruments, a touchstone rock, bamboo pipe, metal pipe, and mortars and pestles. A crucible, tubes, medicinal substances, cloths, sour gruel, black salt, oils, sour substances, salts, alkali, poisonous substances, and narcotics.

7. Sixteen cm (nine *aṅgulas*) high, twenty-eight cm (sixteen *aṅgulas*) long, sixteen cm (nine *aṅgulas*) wide, and measuring ten-and-a half cm (six *aṅgulas*) deep. The edge measures three-and-a-half cm (two *aṅgulas*) thick. The pestle is smooth and is half the height of the vessel and with an eighteen cm (ten *aṅgula*) base. This instrument is the mortar (*khalva*).
8. Eighteen cm (ten *aṅgulas*) high, twenty-eight cm (sixteen *aṅgulas*) long, with a width of eighteen cm (ten *aṅgulas*) and measuring twelve cm (seven *aṅgulas*) deep with half the width of the thumb. The edge is three-and-a-half cm (two *aṅgulas*) wide and smooth. The pestle is like a half moon and measures twenty-one cm (twelve *aṅgulas*). This is called mortar (*khalva*); its use brings perfection.[8]
9. In an earthenware vessel, in an iron vessel, or in one made of magnetic iron[9]—elsewhere it is stated, mercury processed in a mortar made of magnetic iron—will develop ten million qualities (*koṭiguṇa*).[10]

[8] RRS 9.81.
[9] RArṇ 4.59.
[10] RRS. 9.87, ĀK 1.26.13.

10. In this, two hundred and forty grams (five *palas*) of mercury should be triturated for the purpose of purification. In other mortars, mix with whatever is suitable.[11]
11. The mortar is round, twenty-one cm (twelve *aṅgulas*) wide, and seven cm (four *aṅgulas*) deep. The pestle should be fourteen cm (eight *aṅgulas*) long.
12. That which is wrought from metal is called heated mortar (*taptakhalva*). Kindle the fire with chaff and female goat dung, then cast it on the ground. The mortar standing upon this is called *taptakhalva*.[12]
13. A very strong pestle, seven cm (four *aṅgulas*) at the top, is made of any metal, stone, or magnetic iron.[13]
14. Made of metal, it is sixteen cm (nine *aṅgulas*) high and ten-and-a-half cm (six *aṅgulas*) in depth and has a fourteen cm (eight *aṅgula*) pestle. This is called *taptakhalva*.[14]

15–16. In an iron mortar consisting of four parts and a base of eighteen cm (ten *aṅgulas*) another strong mortar is made. Stone slab, clay, or bowls variously made of metal, copper, or clay.

17. Fill the inside of a vessel halfway with liquid. Fix a stick across the opening and hang mercury from a thread and steam it without it touching the bottom. This is called the swing apparatus (*dolāyantra*).[15]
18. Fill the inside of a vessel halfway with liquid. Make an indentation on both sides of the jar, carefully place a stick in them, tie a mercury bundle in the middle, and steam it like that. This is called the swing apparatus (*dolāyantra*).[16]

19–21. Take the *kinnara* apparatus (*kinnarayantra*),[17] smear the bottom with the herbal mixture. Put mercury poured into sal ammoniac inside of the apparatus. Above the mercury is a shallow plate smeared firmly with earth. Fill it with salt and having carefully covered the opening, place it on top of the oven, and kindle a

[11] RCūM 5.8.
[12] RMañj 1.30, RRĀ Vkh 2.46, RCint 3.83.
[13] RArṇ 4.60.
[14] RRS 9.85.
[15] RArṇ 4.7, ĀK 1.26.100.
[16] RRS 9.3–4, MuA 2.3.2.14.
[17] See Hellwig 2009: 202 for a speculative diagram.

strong fire below. Remove this after three hours (*yāma*) and let the mercury become cool.[18]

22. Make a spherical vessel of earth [made of two halves] and fill the bottom of the vessel with a fourth part [of mercury?] and crown flower blossom(?).[19] Knead this a little and make it into the shape of a ball with the hands. After water has been placed on top, carefully seal the seam with clay. Cook it for six hours (two *praharas*) on an even fire. This will raise it up. This describes the enclosed vessel, which is called the hour-glass drum apparatus (*ḍamaruyantra*).
23. Place an earthenware dish face-down over the earthenware apparatus place and close it. This is called hour-glass drum apparatus (*ḍamaruyantra)*, and it is suitable for the calcination of mercury.[20]
24. Place a crucible above an earthenware dish and close it in the same way. Pour water into the elevated dish. Then, light a fire below. This is the wisdom holder's apparatus (*vidyādharayantra*), used to extract cinnabar.

25–27. Fourteen cm (eight *aṅgula*s) wide, eighteen cm (ten *aṅgula*s) long, the water basin under the neck is seven cm (four *aṅgula*s) high. The mouth of this vessel rests on top of the lower vessel below. The mouth has a height that extends twenty-eight cm (sixteen *aṅgula*s). Smear the two sides with buffalo milk, powdered iron rust, and sugar cane juice. Let it dry. Fill the tank with water and place this on an oven. This is called the distillation apparatus (*pātanayantra*).[21]

28. Now, a wise person should carefully carry out the condensation (*adhaḥpāta*) of the (substance) that has been smeared and fixed onto the inner surface of the upper vessel into water with burning cow dung.[22]
29. [A device] in which quicksilver that is positioned over a lamp situated on a base made of clay inside a turtle

[18] MuA 2.6.2.21–23.
[19] The term translated as "crown flower blossom" here is *arkakara*. The translation is uncertain.
[20] RRS 9.57.
[21] RRS 9.6–8.
[22] See RCūM 5.22–23, RCint 3.24–27, RRS 9.9, and Hellwig 2009: 41–44.

apparatus (*kacchapayantra*) drips down, is called lamp apparatus (*dīpikāyantra*).[23]

30. There are many kinds of alchemical apparatuses, such as the "perfected essence" (*siddhisāra-*), "womb essence" (*garbhasāra-*), "the highest bliss manifestation" (*paramānandamūrtija-*), the two devices known as "earth" (*patāla-*) and "sky" (*ākāśa-*), "auspicious dyeing" (*śrīrañja-*) and "the merciful" (*karuṇākara*), "water" (*vāruṇī-*), "the heavenly body" (*jyotis-*), and so on [apparatus].

31. Place a deep bowl filled with paste on an oven. Because drops of liquid essence drip down from a tube on its side, this device is said to be called the perfected essence apparatus (*siddhasārayantra*). The paste is produced from herbs for thickening (*mūrchana*), calcination (*māraṇa*), and solidifying (*bandhana*).

32. Similarly, use a deep vessel with a lid containing holes. Cover the inside vessel with a small lid in the middle. When vapor drops fall it is called "womb essence" (*garbhasāra*).

33. A long tube in the side goes into another vessel, from which distillate falls. This is called "the highest bliss manifestation apparatus" (*paramānandamūrtijayantra*).

34. Fill a bowl with holes with an oily substance and place it on another bowl in a pit. A fire is placed above. The apparatus is called "earth apparatus" (*pātālayantra*).

35–36. On a fire, fit half of a jar face-to-face with an overturned bowl from above. Tie a cloth around the mouth of the vessel. Powdered medicine placed in a cloth falls down into another vessel made of glass and so on when it becomes fluid. This is called sky apparatus (*ākāśayantra*). The spout of the glass bottle is pierced and covered with a cloth.

37–40. Place a very deep and attractive vessel made of earth or glass into a cauldron filled with dyeing liquid. Having raised a tripod with small holes, throw sulfur powder there and heat [stirring] with a small stick until hot. Once burned up, put another similar vessel above the first. After covering it, plunge the contents of the vessel into the fluid within [the lower pot]. Within that surface it will smoke and stick. Thus, proceed and drop oil [through] the hole's

[23] See RRS 9.10–12, RRSTīkā 3–5, RCūM 5.30–33, RRS 9.13, and Hellwig 2009: 178–179 on the "turtle device" (*kacchapayantra*) and 112–113 on the "lamp device" (*dīpikāyantra*).

opening into the middle of the vessel. This auspicious dyeing apparatus (*śrīrañjakayantra*) is taught everywhere.

41. Fill a deep vessel with paste, cover, and place it on a fireplace inside a wide-mouthed pot. When the liquids there on its sides are expelled, it is "the merciful" apparatus (*karuṇākarayantra*).

42. Place the opening of both bottles horizontally below the other, kindle the fire. Then, in another vessel the distilled medicinal salt falls. That is the water apparatus (*vāruṇikayantra*).

43. Light a fire below a raised earthen pot with a pierced hole that allows the essences of Palmyra tree and so on to fall. This is called heavenly body apparatus (*jyotiryantra*).

44. Rest a hollow crucible on a fire place and conceal with another vessel. Essence of Palmyra tree and so on falls. This is called "cooking apparatus" (*sādhyayantra*).

45. A water container that holds eighty-one grams (four *prasthas*) and has a mouth seven cm (four *aṅgulas*) wide is called the jar apparatus (*ghaṭayantra*). It is also called stout (*āpyānaka*).[24]

46. When a hollow crucible below and a hollow crucible above and sand in the vessel sticks to the neck, that is the sand apparatus (*saikatayantra*).

47–48. Fill a bowl with sand, place this above another [bowl]. There, cook it gently with a mild fire. Boil with caustic alkali, urine, salt, and fetid salt. Alchemists call this the swan cooking apparatus (*haṃsa pākayantra*).[25]

49–51. Once dry, fill a glass pitcher that is twelve cm (seven *aṅgulas* high), covered with two cm (one *aṅgula*) mud-smeared old cloths, with three parts ingredients. Stand a twenty-one cm (twelve *aṅgulas*) deep vessel on sand, fill this vessel with the remainder, and cover. Cover the vessel opening with a water pitcher. Smear it with earth and cook on a fireplace without burning the grass resting in the tip water pitcher. This is the sand apparatus (*vālukāyantra*). This apparatus is the seat of salt.[26]

52–54cd. Take two eggplant-shaped crucibles, place two tubes measuring twenty-one cm (twelve *aṅgulas*) below the neck on each. Smear

[24] RRS 9.51.
[25] RRS 9.31–32.
[26] RRS 9.33–35.

the junction with earth. Then, place quicksilver in one and sulfur powder in the other. Stop up the openings of both crucibles and move to a sand apparatus (*vālukāyantra*). Heat them and set the sulfur below on fire. This is known as a balance apparatus (*tulāyantra*). Holes and tubes in the four directions allow smoke to leave.[27]

54–57. Prepare a round hole and place a vessel there. Put down a brick in the middle of a suitable hole. Raise a pot two cm (one *aṅgula*) around the hole. Pour mercury in the middle and place a cloth on the opening of the hole. Put sulfur there and stop up the opening of the vessel. Seal the vessel and the pot around the seam with earth. When cow-dung is used it is called the pigeon vessel (*kapotapuṭa*), not otherwise.[28] This is the brick apparatus (*iṣṭikāyantra*). Digest sulfur with this.

58–60. Make a hemispherical bowl, seven cm (four *aṅgula*s) tall and measuring five cm (three *aṅgula*s) across, and dry it in the shade. Grind twenty parts salt and one part bdellium in water. Now, smear the inside of the bowl with this and stop up the opening. This is the womb apparatus (*garbhayantra*). Its purpose is the digestion of paste.

61. After combining and spreading clay in the middle of a large bowl or tree-root water basin, throw quicksilver in the middle. The purpose of this [stout turtle apparatus (*sthūlakūrmayantra*)] is to assimilate mercury, sulfur, and so on.

62–65. Collect a strong, wide water-filled vessel. Place within this bowl a strong, large new one. Above this, place the objects for the calcination of mercury. Place mercury into the opening. Above those objects, place a magnetic iron cup or separated flat, thin leaves of clay. Placing the spout below from above, smear the combination of crushed chalk, camphor, water, and boiled grain. Open from above with a raised piece of wood. Fill the bowl with air and charcoal. [This is the dolphin apparatus (*jalakūrmayantra*).]

66–69ab. Now, this is certainly called dolphin-class apparatus (*jalakūrmaprakārayantra*). Fill a lower vessel measuring

[27] RCūM 5.63–66.

[28] Described in RRS 10.57: *yatpuṭaṃ dīyate bhūmāv aṣṭasaṃkhyair vanopalaiḥ | baddhvā sūtārkabhasmārthaṃ kapotapuṭamucyate ||*

twenty-one cm (twelve *aṅgula*s) long and twenty-eight cm with four pegs attached at the opening with water. Bury it in the middle of the ground. Set a dish on the opening's four pegs from above. Below, drown a shallow dish in a dish of water. This dolphin apparatus is used for the procedure of assimilation.

69. Place *gajavelli*[29] in a jar, spread it below the opening, and cover as appropriate. This is called the water and fire apparatus (*somānalayantra*).

70–71. Place a wide jug in the middle of a dish filled with water. Above that place quicksilver into special salt (*viḍa*)[30] in an earthen bowl. Cover with a light metal bowl and smear the seam with salt and earth. Fill the jug with with charcoal made of cutch tree.[31]

72. This device for the purpose of binding mercury has water and fire inside. Place a large-mouthed earthenware vessel in water on an oven.

73. Fire above, water below, and mercurial compound in the middle. This (device) is called water and fire (*somānala*). Digest a quarter portion of mica and pyrite essences that are melted together in it.[32]

74. [A device made of] ash of soil, burnt chaff, and burnt termite mound clay,[33] fetid salt, charcoal, iron rust together is not split even with a diamond.

75. Six parts of roasted charcoal, one part black earth, fetid salt, charcoal, and iron rust. This is called the diamond crucible (*vajramūṣā*).[34]

76. Equal parts of burnt chaff and fabric, and four parts earth combined with earth stone, is called diamond crucible (*vajramūṣā*).

77–78. Take one part of iron rust, double that of roasted chaff, and join with three parts earth. Add any base metal. Crush all these together. The clay becomes diamond-like. An adept makes three types of crucibles with this clay.

[29] I was not able to identify *gajavelli*.

[30] I have translated *viḍa* as "special salt" here, since there is more than one kind of *viḍa*. On the various preparations that go under the name of *viḍa* and are meant to help mercury digest, i.e., assimilate the substances fed to it, see Hellwig 2009: 446–448.

[31] This charcoal is of the *kokila* type, meaning that it has been extinguished without water. See Sauthoff, this volume. RCūM 16–24ab.

[32] RRS 9.26.

[33] See Smith and Carri 1994: 224n7 on *valmīka* as termite mound rather than an anthill.

[34] RArṇ 4.36.

79. Iron rust, roasted chaff, and shell powder together with gravel in equal shares and an equal part earth. Bind the crucible. Through firing it will become as hard as a diamond.

80. One part brick powder and an equal share clay. Through firing it will become as hard as a diamond.

81. Crush three parts earth, two parts salt, and one part roasted chaff, precious stone, and so on, a half part iron rust. This is a diamond crucible (*vajramūṣā*). It brings about drops of essence.[35]

82ab. Circular and made of cow dung, it is called diamond crucible (*vajramūṣā*).[36]

82cd–85ab. Mix one part termite mound clay, two parts chaff ash, one part of iron rust, and one part rock crystal and an equal amount of human hair. Crush this firmly with goat's milk for six hours. With this, a crucible is made into a spherical bowl. Now, having dried mercury and placed it inside, the joints are sealed with the paste. This is known as a diamond crucible (*vajramūṣā*). It is properly used in the killing of mercury. The chaff and so on should be burnt until it becomes black.

85cd–86. Burn chaff and an equal amount of burnt cloth with a quarter of clay. Crush with a quarter of stone and distillate of veldt grape for a day. Use the crucible known as diamond (crucible) to bind mercury.[37]

87–88. Cook roasted red chalk, roasted chaff, roasted termite mound clay, roasted excrement of elephant, and horse until it has gone blackish. Dry and burn coleus leaves and an equal amount of black clay with veldt grape distillate, clay, and turmeric. With this, one should make a kiln, a blow pipe, and a diamond crucible (*vajramūṣā*).[38]

89. Smear a glass or iron cup of seven cm (four *aṅgulas*), nine cm (five *aṅgulas*), or sixteen cm (nine *aṅgulas*) with seven layers of cloth and clay and dry it.

90. *Ācārya's edition skips 90.*

[35] RRS 10.9.
[36] RRĀ Vkh 3.24cd.
[37] RRĀ Vkh 18ef–19ab.
[38] RRSTikā 10.32.5cd–7ab.

91–92. Powder termite mound clay, charcoal, iron rust, and white calcareous spar. Rub all this in equal parts with roasted chaff and water. Smear this onto all the joints of the covers of crucibles and so on.[39]

93. A shaped and fired vessel that is solid for roasting, that is called the roasting crucible (*pakvamūṣā*). Everywhere it is used in roasting.[40]

94. An open and a covered crucible are thought to made in two ways. The open crucible has the shape of a shallow dish (*śarāva*), O Deveśī. [Used for] the removal of liquids, it is praised in medicine.

95–96. One should make a covered crucible that is similar in appearance to cow dung with a lid with a curved top. [Used in the procedures of] leaf coating and fusing [metals and minerals], it is deep and hollow, and it is suitable for potentiating (*sāraṇa*) mercury.

97–98. Having made a tall crucible for the potentiation of mercury and the extraction of essences, make a thick cover. The dish is suitable for burning. Make a crucible of sixty-six cm (one and a half *hasta*) of metal. Cover the crucible on top, smear the joints, and stick a curved tube that looks like cereus into the hole in the enclosed vessel.

99. Two parts sesame ash together with one part brick is called the ash crucible (*bhasmamūṣā*). It is fit for the purification of silver.[41]

100–101. A box-shaped crucible that is deep and wide, and can hold volume with a measure of ten-and-a-half cm (six *aṅgulas*), is called the box crucible (*mañjūṣamūṣā*).[42] Bury the crucible in earth and place a cover upon it.[43]

102. A closed spherical [vessel] filled with substances for enclosed roasting is called the globe crucible (*golamūṣā*). It quickly produces liquids.[44]

[39] RRĀ Vkh 3.19cd–21ab.
[40] RRS 10.27, RPSudh 10.25, ĀK 1.26.173.
[41] RArṇ 4.42, ĀK 1.26.184cd–185ab.
[42] *maṇḍūkākamūṣā* in ĀK, RCūM, and RRS.
[43] ĀK 1.26.176cd–177, RCūM 5.125, RRS 10.30.
[44] The reading here is somewhat uncertain and differs from parallel passages in the other works, which seem to describe the removal of liquids. RRS 10.28, RcūM 5.123, ĀK 1.26.174. See Hellwig 2009: 151.

103. Equal parts roasted chaff, iron rust, red ochre, potters clay, white clay, chalk. It is called the little dried-in-shade crucible (*chāyāśuṣkamūṣikā*).

104. Termite mound clay, smoke (=soot?), gold, chalk, and salt, these are called the five earths in alchemical operations.[45]

105. Earthworm cast, clay, and hard iron rust, chaff of rice, and grain are roasted with great care.

106–110. Black, red, yellow, and white colored clay. The first is thought to be best, the last medium, and the others medium. Clay mixed with roasted chaff of grain is praised for [making] a kiln or made into a blow pipe, O Surasundarī. Crush roasted mud,[46] roasted chaff, roasted termite mound clay, roasted goat and horse excrement, and roasted black earth together with leaves of malabar nut, and termite mound clay with leadwort (?) water until it has gone smooth. With that rub and fix the blow pipe and kiln.[47] Smooth, slimy, heavy, black clay are all recommended, excluding yellow or that with its qualities, or joined with sand and so on.

111–112. Clay with roasted chaff, charcoal, horse droppings crushed with an iron stick. That is universal for a crucible. Crush earth, potter's clay, or termite mound clay with chalk, salt, iron rust, and buffalo milk in fire. Clay becomes able to withstand the heat of fire.

113–114. Crush boiling syrup-like gum arabic, old fine rust powder mixed with jaggery and lime. This is called water-clay (*toyamṛttikā*). It is difficult to be breached with water.

115–118. Crush earth prepared with clay, cotton, and iron rust, or with lac, clay powder salt, jaggery, mung bean, and water and so on, with milk of a ewe grazing on old cotton. Smear a glass bottle with more earth crushed with rice water. These two earths are the best. Crush green sulfate powder, iron rust, borax, ammonium chloride, the five salts, and clay with lime juice. This

[45] RCint 6.8, RRS 1.249, ĀK 2.215. The term translated as "alchemical operations" is *rasakarman*, which refers to the program of procedures applied to mercury.

[46] The substance *gāra*, translated here as "mud," is somewhat unclear, as it is defined in different ways in different texts. See Hellwig 2009: 141–142. Ācārya's edition of the *Rasakāmadhenu* does not contain a definition of *gāra*. However, the edition by Mishra and Sharma (1992) does: 1.1.169 reads: "*gāra* is decayed clay that was in water for a long time" (*jale ciraṃ śīrṇamṛttikā gāram*).

[47] RArṇ 4.30–35.

foam-like clay is recommended for application to the hourglass drum[48] apparatus (*ḍamaruyantra*).

119. Rub black potter's earth, yellow earth, coarsely ground oats, human hair, paper (?), and grain with grape, jaggery and so on, vinegar. This is called the diamond crucible clay (*vajramūṣāmṛd*).

120–121. Chalk, salt, red ochre, fragrant earth, and brick powder. These are called the five soils and bring wealth and success. Smearing and enveloping with these brings agreeable wealth.

122–127. Connect two parts weaver's beam tree wood alkali with one part brick and portions of earth; for the purification of silver, this is the most excellent dye. As a mixture of red substances, an immersion in red substances, smearing with red substances, it is suitable in all purifications. As a mixture of white substances, an immersion in white substances, smearing with white substances, it is suitable for purifications of white substances. The group of salts mixed with earth are recommended in the procedure of digestion. Perform the evaporation process by application of red substances. Smear the crucible with aconite, borax, and jequirity beans (*guñja*) or with charcoal if preparing it in an open crucible. Having placed it into that crucible, a wise person melts the substance in this crucible. O Devī, this smearing with red ochre, rock-salt, and green vitriol [is applied] in the color roasting process.[49]

128. Breast milk, borax, antimony, flame lily, butterfly pea, leadwort, oleander, Indian sarsaparilla, euphorbia, aconite. Smear these on the crucible. It is a great drug[50] for the purpose of binding (mercury).

129. Chalk, salt, mung bean flour, jaggery, bdellium, flax, and stone powder are revered for the process of coating with clay and so on.

130–131. Coat the joint diligently with clay and also with cloth. Apply alkali and borax on the joint of the two crucible covers. Place it on a fire. The joints will strongly fuse.

[48] The name of the device refers to the shape of the *ḍamaru*, a small drum that has the shape of an hour-glass.

[49] RArṇ 4.43–48.

[50] The great drugs (*mahauṣadha*) are a category of auxiliary herbs used in the processing of mercury. On these categorizations, see Chapter 8 by Dagmar Wujastyk in this volume.

132. White stones, blackened chaff, charcoal, hemp fibers, animal droppings, iron rust, and black earth are applied as clay to a bottle.[51]

The coating [for bottles] is made from water and iron[52] with the addition of chalk and salt.

133–138. Two parts chaff, one part cloth fragment, and clay. Arrange in three layers, add water, and rub. Take the equivalent amount of human hair and grind it a bit until this makes a clay ball that looks and feels like wax. Do this diligently so that it docs not dry out. After seven days, a physician should gradually apply the clay in a mixture as a coating to a bottle and so on and to an apparatus. Make an equivalent to that using water, gum arabic resin. Equal to this, iron rust. It resembles purified collyrium. Prepare the seal of the fluid apparatus (*vāriyantra*), which is difficult to obtain for the purpose of perfection. Here, it never breaks because of fire or water. Mercury does not get lost even through the force of the fire of dissolution.

139–141. Mix forty grams (two *kuḍava*) of flax oil with eighty-one grams (a *prastha*) of water. Prepare the remaining oil on fire. In the middle of that burn birch leaves equal to the ointment. Join forty-eight grams (one *pala*) of the remaining dust, grain, and so on, and forty-eight grams (one *pala*) iron rust resembling purified collyrium. Rub it with lime juice and apply it. This is its purpose.

142–144. Use one part alkaline salt, two parts calcined gold, five parts zinc, and six parts washed clay. Crush this with willow juice and form a hemispherical bowl with it. Form a crucible to truly restrain mercury. It will not get lost because of being in fire and becomes one whose wings are cut. A glass bottle is the second, an iron hemispherical bowl the third.

145–146. Use equal parts flame of the forest ash, chaff flower, and barley alkali, sour gruel, socal salt, iron sulfate, cuttlefish bone, rocksalt, black salt, borax, sal ammoniac, camphor, pyrites, and the sap of common milk hedge and crown flower and smear onto a crucible, O Deveśī.[53]

[51] RRS 10.8.

[52] The text reads *kīlālāyaḥkṛto lepaḥ* here. *kīlāla* could be water, but could also be honey, or even blood. The passage also occurs twice in the *Rasendracintāmaṇi*: in 2.18 and in 1.151.

[53] RAṛn 87–89ab.

147–151. [Use one part each] of the sap of common milk hedge and crown flower, mulberry, guggul, and two parts rock-salt. A skillful person should crush and wrap this paste around a flour cake, O Sundarī. Make a new, very firm crucible of forty-eight grams (one *bilva*). Put mercury in the middle of the crucible with salt placed above and below. Coat the crucible using roasted shell powder and so on. Having firmly sealed the joints of the crucible with salt and clay, apply a coat of honey[54] and let it dry in the shade. This is called restrained binding (*nigaḍabandha*). It is not even told to a son.

152–156. Sesame oil, crown flower sap, date tree, blue water leaf[55] are a suitable binding agent. Crow feces, mulberry tree, and blue water leaf are a suitable binding agent. Bakuchi, mulberry tree, milk hedge and crown flower sap, rock-salt, devil's tongue, and crow feces are praised as the unparalleled binding agent. Crush even parts salt, borax, alkali, shilajit, orpiment, sulfur, curly dock,[56] pyrite, and cinnabar joined with milk hedge and crown flower sap. This is the very best binding agent. Envelop the powder with a binding agent [made of] one of these. As before, having made mercury lame[57] in a metal crucible, let it penetrate.

157–158. Join magnetic iron and iron powder with hog's blood. Then, having placed all of these together, beat them with a solid stroke from dawn to dusk. This is called the forceful seal (*haṭhamudrā*), worshiped by all perfected beings.

159–160. Give twelve grams (three *ṭaṅkas*) magnetic iron, thirty-six grams iron rust joined simultaneously with the blood of a man. With a pestle crush all for twelve hours. This is known as forceful seal (*haṭhamudrā*), worshiped by all perfected beings.

161–164. Join and crush two hundred eighty-eight grams (six *palas*) iron slag and strain it through a cloth into a mortar. Mix goat and ram heart and flesh. Take it and pound it with a steady hand

[54] *Sudhā*, translated as honey here, may be milk hedge instead.
[55] Chopra, Nayar, Chopra, 1956: 137.
[56] Chopra, Nayar, Chopra, 1956: 216.
[57] Hellwig 2009: 200–202. This is mercury that has been calcined several times.

having added it to the powder until it is like bees wax. This will be an effective seal. There are said to be other seals that are prepared though various combinations. Quickly prepare glass, borax, iron rust, and brick with the sap of banyan tree, crown flower, and Indian fig for one day. This becomes a forceful (seal).

165. Grind forty-eight grams each Indian fig wood, banyan tree sap, lac, three hundred thirty-six grams magnetic iron mixed with flax fruit oil. In the method for the digestion of quicksilver this is called the wax seal (*madanamudrā*).

166–167. Gradually heat water in four metal pots. Cook this well in a pan by adding an increasing amount until the wax that was on the bottom floats on the water. The wax is declared the best in the coating onto water apparatuses (*jalayantra*) and so on.

168–179. A wise person puts two layers of thin old cloth on a square glass bottle measuring half or one and half. Stretch a fourth portion of disintegrating (cloth?) in the middle, immerse it in muddy clay, and apply it to the glass bottle. On this smear a measure of barley and again with solid clay. Put one piece of cloth on the middle of the bottle's hole. After placing this on the bottle tube, evenly distribute clay on all sides. Tie another cloth piece below this on the glass bottle. Dry this in heat, then reapply clay. Use another two fragments of cloth. Join the last to the third rag with clay and dry. Seal the powder of mercury and so on inside with great care. Cover both the neck and tube with ashes. Place five or six charcoal pieces in the neck and burn until very bright. Use the mouth to blow air through a metal or bamboo tube until it drips down the neck of the glass bottle. With tongs, carefully and evenly obstruct the joint and seal the melted glass with *gorakṣa*. Gently remove the charcoal with tongs. Then, once the bottle is sufficiently cooled, take it and as before remove the four layers of rag and clay. Smear with minerals, chaff, or clay, avoiding the hole. Let this dry. Hold the apparatus, *cakra*, and so on, and perform the assimilation of metals. Thus, with smokeless cooking, the result is beyond ten million.

Bibliography

Ānandakanda. Edited by S. V. Radhakrishna Sastri. Srirangam: Sri Vilasam Press. 1952. Digital Corpus of Sanskrit e-text. Input by Oliver Hellwig.

Chopra, R. N., S. L. Nayar, and I. C. Chopra. 1956. *Glossary of Indian Medicinal Plants*. New Delhi: Council of Scientific & Industrial Research

Dole, Vilas, and Prakash Paranjpe. 2016. *A Text Book of Rasashastra*. Delhi: Chaukhamba Sanskrit Pratisthan

Golovkova, Anna A. 2020. "The Forgotten Consort: The Goddess and Kāmadeva in the Early Worship of Tripurasundarī." *International Journal of Hindu Studies* 24 (1, April): 87–106.

Hellwig, Oliver. 2009. *Wörterbuch der Mittelalterlichen Indischen Alchemie*. Groningen: Barkhuis.

Meulenbeld, G. Jan. 2000a. *A History of Indian Medical Literature*. Volume IIA. Groningen: Egbert Forsten.

Meulenbeld, G. Jan. 2000b. *A History of Indian Medical Literature*. Volume IIV. Groningen: Egbert Forsten.

Mugdhāvabodhinī of Caturbhuja Miśra. Unedited. Digital Corpus of Sanskrit e-text. Input by Oliver Hellwig.

Rasakāmadhenu of Cūḍāmaṇi. 1988 [1925]. Edited by Yādavajī Trivikramajī Ācārya. Varanasi: Caukhambhā Oriyanṭāliyā.

Rasakāmadhenu of Cūḍāmaṇi. 1992. Edited by Jīvarāma Kālidāsa Śāstri. Varanasi: Caukhambhā Oriyanṭāliyā.

Rasakāmadhenu of Cūḍāmaṇi. 1992. Edited by Gularaj Sharma Mishra and Santosh Kumar Sharma. Varanasi: Caukhambha Orientalia.

Rasakāmadhenu of Gularaj Sharma Mishra and Santosh Kumar Sharma. 1992. Varanasi: Caukhambha Orientalia. Digital Corpus of Sanskrit e-text. Input by Oliver Hellwig.

Rasamañjarī of Śālīnātha. 1995. Edited by Siddhinandan Mishra. Varanasi: Chaukhambha Orientalia. Digital Corpus of Sanskrit e-text. Input by Oliver Hellwig.

Rasaratnasamuccaya of Vāgbhaṭa. 1890. Edited by Kṛṣṇaravasarman Vinayaka Bapta. Poona. Digital Corpus of Sanskrit e-text. Input by Wieslaw Mical, Som Dev Vasudeva, Anne Glazier; Oliver Hellwig.

Rasaratnasamuccayabodhinī of Āśubodhavidyābhūṣaṇa and Nityabhodhavidyāratna. Unedited. Digital Corpus of Sanskrit e-text. Input by Oliver Hellwig.

Rasaratnasamuccayaṭīikā. Unedited. Digital Corpus of Sanskrit e-text. Input by Oliver Hellwig.

Rasaratnākara of Nitya Nātha Siddha. 1939. Edited by Vaidya Jādavjī Tricumjī Ācārya. Benares: Caukhamba Sanskrit Office. Digital Corpus of Sanskrit e-text. Input by Oliver Hellwig.

Rasārṇava. 1910. Edited by Praphulla Chandra Ray. Calcutta: Hitavrata Chattopadhyaya. Digital Corpus of Sanskrit e-text. Input by Oliver Hellwig.

Rasendracūḍāmaṇi of Somadeva. 1999. Edited by Siddhinandan Miśra. Benares: Chaukhamba Orientalia. Digital Corpus of Sanskrit e-text. Input by Oliver Hellwig.

Smith, Frederick M., and S. J. Carri. 1994. "The Identity and Significance of the Valmīkavapā in the Vedic Ritual." *Indo-Iranian Journal* 37 (3): 201–231.

White, David Gordon. 1996. *The Alchemical Body*. Chicago: Chicago University Press

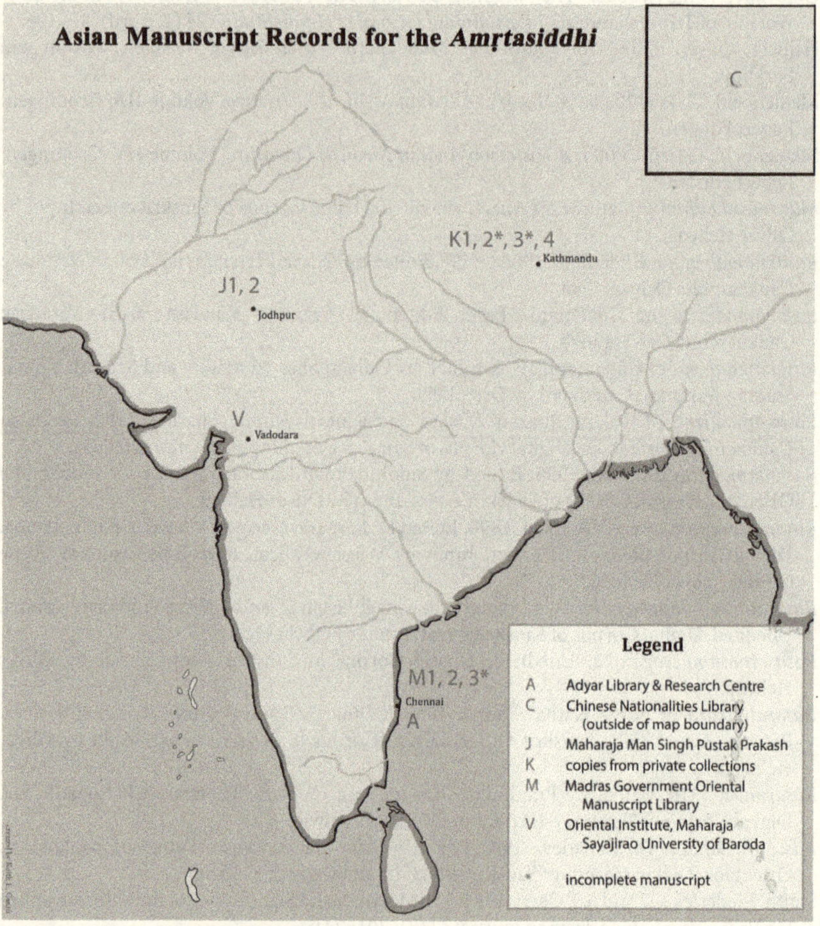

Map 10.1 Asian manuscript records for the *Amṛtasiddhi*.

10

The *Amṛtasiddhi*

Haṭhayoga's Flash in the Alchemical Pan

James Mallinson

The *Amṛtasiddhi* or "Attainment of Immortality" is a Vajrayāna text in Sanskrit on physical yoga that was composed most probably in the eleventh century in the Deccan region of India.[1] It is the first text to teach any of the practices and principles that come to be categorized as *haṭhayoga* in subsequent Sanskrit texts.

The yoga of the *Amṛtasiddhi* uses a sequence of physical methods to control the breath (and hence the mind), raising it up the body's central channel, thereby also raising three of the bodily elements situated at the base of the channel: the sun, fire, and *rajas* (the female generative principle). When these unite in the head with the moon and *bindu* (the male generative principle) the yogi attains *jīvanmukti*, liberation while living.

The *Amṛtasiddhi* does not teach external alchemy, but much of its teaching on yoga is framed in alchemical metaphors and uses alchemical terminology.[2] This terminology is examined below, followed by an overview of alchemy's role—or lack of—in yoga after the composition of the *Amṛtasiddhi*. There then follows a translation of chapters 11–14 of the text, which teach its yoga's physical techniques and sequence of practice.

Prior to the composition of the *Amṛtasiddhi*, yoga texts and traditions teach that the body is to be subjugated, or even mortified with severe

The research for this article was carried out as part of the Hatha Yoga Project (hyp.soas.ac.uk). This project has received funding from the European Research Council (ERC) under the European Union's Horizon 2020 research and innovation program (grant agreement no. 647963).

[1] For detailed information on the *Amṛtasiddhi* see the critical edition and translation of the text by Mallinson and Szántó. Much of this chapter is drawn from the section on alchemical metaphors in the edition's introduction (Mallinson and Szántó 2022: 20–22).

[2] Cf. the *Vimalaprabhā* commentary on the *Laghukālacakratantra*, which uses alchemical analogies to explain yogic practices, such as at 5.125 in which the control of semen is compared with the alchemical binding of mercury. Unlike those of the *Amṛtasiddhi*, however, the yogic techniques taught in the *Vimalaprabhā* are not described in alchemical terms.

James Mallinson, *The* Amṛtasiddhi In: *Indian Alchemy*. Edited by: Dagmar Wujastyk, Oxford University Press.
 DOI: 10.1093/9780197805831.003.0011

austerities. The *Amṛtasiddhi*'s alchemy-framed body-positive yoga has no precedents in Indian traditions, but its metaphysics and metaphors do echo treatments of inner alchemy, or *neidan*, found in Chinese texts dating from two to three centuries earlier. These texts are themselves innovations in the Chinese alchemical tradition, whose practices had previously been only external.[3] The physical methods of the *Amṛtasiddhi*'s yoga, however, do not have any known precedents in India, China, or elsewhere.

Alchemical Terminology in the *Amṛtasiddhi*

The Three Practices and Other Alchemical Transformations

The names of the three primary techniques of the *Amṛtasiddhi*'s yoga, *mahāmudrā* "the Great Seal," *mahābandha* "the Great Lock," and *mahāvedha* "the Great Piercing," are alchemical terms, which are here applied to physical yogic techniques.

Mahāmudrā, "the Great Seal"
For this technique the yogi, sitting on the ground with the buttocks raised slightly on a mat, places the left heel at the perineum, extends the right leg, holds the right foot with the hands, puts the chin on the chest, and inhales deeply. The name *mahāmudrā* implies that the breath is being sealed within the body. In alchemical texts the joint between the two halves of a closed crucible (a *saṃpuṭa*, on which see below) may be closed off with a seal (*mudrā*) made from a paste of ash or mud.[4] *Rasacintāmaṇi* 3.87 refers to a seal of this kind called a *mahāmudrā*.[5]

Mahābandha, "the Great Lock"
Remaining in *mahāmudrā*, the yogi now applies *bandha*s on (i.e., performs contractions of) the regions of the perineum and throat, locking the breath in the abdomen. *Bandha* (or *bandhana*) is an alchemical *saṃskāra* and a process in which mercury is stabilized and made more resistant to heat.[6] In the *Rasaratnasamuccaya*, mercury is called *mahābandha* when it is heated

[3] See, e.g., Pregadio 2019: 35–67 and Yang 2023.
[4] E.g., *Rasaprakāśasudhākara* 1.83 (Hellwig 2009: 172).
[5] Hellwig 2009: 279.
[6] Hellwig 2009: 68 and Dagmar Wujastyk in Chapters 1 and especially 8 in this volume on the subject of alchemical *saṃskāra*.

with gold or silver in such a way that the two reagents combine to create a solid but friable sphere.[7]

From the alchemical usage of *bandha* and related terms, one may infer that the *Amṛtasiddhi*'s *mahābandha* is intended to stabilize or "fix" the breath once it has been sealed in the abdomen. Elsewhere in the text, at verse 7.7, various states of *bindu* (male generative fluid) are mentioned, including *baddha*, "fixed" (for more on this verse, see below).

Mahāvedha, "the Great Piercing"

The yogi now squats down, places his hands on the ground in front of him, lifts up his buttocks, and drops them onto the heels, thereby sending the compressed breath up the central channel. *Vedha* is a common alchemical *saṃskāra* and denotes the transformatory amalgamation of two reagents; in the yoga practice of the *Amṛtasiddhi* it means both the piercing of the central channel with the breath and, more broadly, the transmutation of the body.[8] In addition to the names of these three practices, five other terms for alchemical transformations are used in the *Amṛtasiddhi* to describe the effects of yogic practice.

Cāraṇa, "Activation"

In alchemical texts *cāraṇa* (literally "making move") is the swallowing of a substance by mercury before its assimilation (*jāraṇa*, on which see below).[9] The word is found at *Amṛtasiddhi* 11.6 and 14.13. In the first instance its alchemical meaning seems inappropriate—there is no reason why *bindu* (male generative fluid) and *nāda* (internal resonance) should be swallowed, so it has been translated here with its literal meaning of "activation." In 14.13 the object of *cāraṇa* is *rasa* (the products of digestion), and it is said to happen "continuously in the channels of the body," so here the alchemical meaning is perhaps apt, but the literal meaning is even more so, so it has been translated as "making move."

Jāraṇa, "Digestion"

In alchemical texts the term *jāraṇa* (sometimes found as the feminine *jāraṇā*) refers to the assimilation of a substance by mercury.[10] It is used at

[7] *Rasaratnasamuccaya* 11.92 (Hellwig 2009: 229).
[8] *Vedha* is also found with this dual sense in the *Kālacakratantra*, on which see Fenner 1979: 72.
[9] Hellwig 2009: 174.
[10] Hellwig 2009: 102

Amṛtasiddhi 11.6 to denote the assimilation of impurity, and at 14.12 to denote the assimilation of *rasa*, that is, the lesser products of digestion, thereby bringing about an abundance of the chief *rasa*, that is, *bindu* and/or *rajas* (the male and female generative fluids, both of which exist in the body of the male practitioner).

Māraṇa, "Killing"

In alchemical texts *māraṇa* ("killing") involves heating a substance and thereby changing its state, usually through calcination or oxidation, so that it becomes inert.[11] In the *Amṛtasiddhi māraṇa* and *mriyate* (a finite verb also from the root *mṛ*, "die") are used at 7.24 and 14.14 respectively to denote the stilling or stopping of the breath, *bindu*, mind, and other bodily elements (*tattvas*).[12]

Mūrcchana, "Thickening," and *Laya*, "Dissolution"

Amṛtasiddhi 7.7, which has multiple parallels in other texts including the *Hevajratantra* (its probable source),[13] describes the result of transformations of *bindu* in terms that in alchemical texts are used for mercury: it is *mūrcchita* "thickened," *baddha* "fixed" (see *mahābandha* above), and *līna* "dissolved."

"Thickened, it removes disease; bound, it makes one a Sky-Rover; absorbed, it brings about all supernatural powers and unmoving it bestows liberation."[14]

Other Alchemical Terminology in the *Amṛtasiddhi*

Rasa

In the *Amṛtasiddhi*'s fourteenth chapter (*viveka*), the term *rasa* is used in the sense of the nourishing fluid produced from good digestion, but its conjunction with some of the alchemical terms noted above (*jāraṇa* and *cāraṇa*)

[11] Hellwig 2009: 238–240.

[12] *Amṛtasiddhi* 7.23, 14.14.

[13] See the testimonia in the apparatus of the critical edition for details.

[14] See the second entry under *Kalyāṇakāraka* in http://ayuryog.org/alchemytimeline (consulted on June 28, 2021) for a parallel verse and references to variations of the verse in alchemical, medical, and yoga literature.

Figure 10.1 Alchemist Siddha. Cave 29 (Gaur Lena) Panhale Kaji, ca. thirteenth century. Photograph James Mallinson 2016.

strongly suggests that the author was consciously seeking to echo its use in alchemical texts, where it refers to mercury.

Saṃpuṭa, "Bipartite Crucible"

In the *Amṛtasiddhi*, the state attained by *mahābandha*, the Great Lock, is called *saṃpuṭa* (12.16, 20.8). In Sanskrit texts on alchemy there are several types of *puṭa* or crucible, bowl-like vessels in which alchemical operations, in particular *māraṇa*, are performed. A *saṃpuṭa* consists of two *puṭa*s joined together to form a sealed crucible for heating reagents without evaporation.[15] In the *mahābandha* practice, the breath is fixed inside the body by performing two constrictions, one at the perineum and one at the throat, and thus the body becomes like a *saṃpuṭa*.

Ghaṭa, "Pot"

In *viveka* 20, which describes the *ghaṭāvasthā* or Pot Stage, the second of the four stages of the *Amṛtasiddhi*'s yoga, it is said to be called *ghaṭa* because

[15] Hellwig 2009: 342. A variety of vessels to be used in a *rasaśālā* may be seen at https://tinyurl.com/tc966jd. The shape of the first depicted, the *svedani* [*sic*] *yantra* is the same as that of a *saṃpuṭa*.

saṃpuṭa yoga is attained in it. The Sanskrit word *ghaṭa* is a very common term for any sort of pot, but may also refer to a type of *puṭa* or alchemical crucible.[16]

Alchemy in Subsequent *Haṭhayoga* Texts and the Broader Yoga Tradition

The *Amṛtasiddhi* does not teach external alchemy, but, as shown above, the language it uses in its teachings on yoga is steeped in alchemical terminology. The redactor of the *Amaraugha*, the first non-Buddhist text to teach a *haṭhayoga* named as such, based that text's description of *haṭhayoga* on the teachings on *mahāmudrā*, *mahābandha*, and *mahāvedha* in the *Amṛtasiddhi* (while adding Kaula Śaiva features). The redactor clearly understood the *Amṛtasiddhi*'s alchemical metaphors, as shown by his use of *maraṇa* and the synonyms *dvipuṭa* and *puṭadvaya* for the *Amṛtasiddhi*'s *saṃpuṭa*: *Amaraugha* 19–21 says that as a result of *mahāmudrā* and *mahābandha* the coiled Kuṇḍalinī becomes straight and attains the state of *maraṇa*, that is, is stilled, when she is in the double crucible (*dvipuṭa*); and verse 29 says that as a result of *mahāvedha* the breath bursts forth from the double crucible (*puṭadvaya*).

Subsequent Sanskrit texts on *haṭhayoga*, however, seem to misunderstand (or perhaps, in some early instances, ignore) the alchemical cast of early *haṭhayoga*. As we have seen above, the name of the *ghaṭa* stage of yoga in the *Amṛtasiddhi* derives from the body's comparison to an alchemical crucible, for which *ghaṭa* is a possible name. In a 1200 CE Vaiṣṇava *haṭhayoga* text called the *Dattātreyayogaśāstra*, however, the *ghaṭa* stage is said (at verse 90) to be named thus because it arises when various pairs unite (*ghaṭate*, derived from the root *ghaṭ* "happen"). More than half a millennium later, two nineteenth-century commentators on *Amaraugha* 21 as found in two recensions of the *Haṭhapradīpikā* improbably take the compound *dvipuṭa*, which in the *Amaraugha* means "double crucible," to refer to the two nostrils.[17]

[16] Hellwig 2009: 18.

[17] Bālakṛṣṇa in his *Yogaprakāśikā ad* ten-chapter *Haṭhapradīpikā* 5.17 and Brahmānanda in his *Jyotsnā ad Haṭhapradīpikā* 3.12.

An exception to this trend toward misunderstanding or ignoring alchemy is found in the *Khecarīvidyā*, which, like the *Amṛtasiddhi*, uses alchemical terminology as part of its description of its yoga methods. Unlike that of the *Amṛtasiddhi*, however, the *Khecarīvidyā*'s alchemy-tinged yoga feels like a trumping of alchemy: echoing the alchemical *mardana saṃskāra* ("trituration") in which substances are rubbed into mercury, at 2.72–79 the yogi is to rub himself with bodily excreta in order to attain various physical *siddhi*s. Elsewhere in the *Khecarīvidyā* (1.68), the physical practice taught in the text is said to give rise to a range of *siddhi*s, including that of mastery of alchemy.

One other stream of the diverse transmission of the *Amṛtasiddhi*'s yoga does preserve a hint of its alchemical flavour. The yoga taught in early Hindi works is close to that of the *Amṛtasiddhi*, and they too couch their teachings in a metaphor of fluid transformation, but it is of distillation, not alchemy.[18]

Some texts of the *haṭhayoga* corpus, in addition to divesting *haṭhayoga* of its original alchemical trappings, have an attitude toward alchemy that is at best ambivalent and sometimes explicitly condemnatory. Thus *Dattātreyayogaśāstra* 52 includes alchemy (*dhātuvāda*) among the obstacles to success in yoga. This ties in with its treatment of other typically Śaiva practices elsewhere, including its allocation of *mantrayoga* to the lowliest type of practitioner (verse 14). *Dattātreyayogaśāstra* 99 and *Śivasaṃhitā* 3.61 say that the yogi can turn iron into gold by smearing it with his urine and/or faeces, a reference to alchemical practice that is likely to be derogatory in the same way as the *Khecarīvidyā*'s *mardana*.[19] The *Yogabīja* (verse 130) says that the *siddhi*s that arise through *rasavāda* are inferior to those which arise through yoga. The thirteenth-century Marathi *Jñāneśvarī* says that through the grace of the guru the yogi can turn iron into gold, which the alchemist cannot.[20]

Despite the divergence between alchemy and yoga in Sanskrit and North Indian textual sources after their brief entanglement in the *Amṛtasiddhi*, in secondary literature the two are often understood to be closely connected. This has led to a notion that success in yoga results in the sublimation of semen, that is, its transformation into *amṛta*, the nectar of immortality,[21] but this is not found in textual treatments of the subject. There are two reasons

[18] See, e.g., Kabīr *Rāga Gauḍī* 20 (Strnad 2013:30). Cf. Caryāgīti 3 (Kvaerne 1977: 81–86).

[19] The alchemical *Rasārṇava*, however, makes a similar claim (12.20ab; 18.28–29. I thank Jason Birch for this reference).

[20] Cf.~*Matsyendrasaṃhitā* 4.85 and *Haṭhatattvakaumudī* 25.17.

[21] White 1996: 45.

for the mistaken conflation of alchemical and yogic traditions. Firstly, an overdependence on sources from South India, where late medieval Tamil Siddha traditions of yoga and alchemy are indeed closely related.[22] And, secondly, confusion over the name yogi, which is used in a wide range of Indian languages as a generic term for a practitioner of any tantric discipline (including alchemy), not just yoga. Thus *yogī*s (often in the vernacular form *jogī*s) described in a wide range of sources from the fourteenth century onward are most often practitioners of alchemy rather than yoga.[23] Some texts, however, do distinguish the two: *Gorakh Bānī Sabadī* 210 mentions a *rasāyanī* and a *jogī*.

Notes on the Text and Translation

The translation is taken from the critical edition and annotated translation by James Mallinson and Péter-Dániel Szántó published in 2022. The edition is based on eleven manuscripts from across the Indian subcontinent and one from Tibet, which is the oldest witness of the text (possibly as old as 1160 CE) and often provides the best readings. Some of the Sanskrit words in the text have not been translated into English. These are: *Bindu*, which literally means "drop," but in the *Amṛtasiddhi* refers to semen; *Nāda*, the internal sounds heard in the course of yogic practice; *Prāṇa*, which is the breath in the upper part of the body; *Apāna*, breath in the lower part of the body; *Siddha*s, adepts of yoga; *Udāna*, breath located in the region of the throat; and *Saṃpuṭa*, bipartite alchemical crucible. Technical terms are capitalized.

[22] See Mallinson 2014: 172n31.

[23] See for example Sastri 1939: 177–179 on Marco Polo and ibid.: 281 on Ibn Battuta; Digby 2025 passim on encounters between yogis and sufis in North India.

Translation of *Amṛtasiddhi*, Chapters 11–14

Chapter 11, The Inquiry into the Great Seal (*mahāmudrā*)

1. That Great Seal (*mahāmudrā*), which is hidden in all the Tantras, is [now] taught. Having obtained it men are worshiped by gods.
2. After performing a host of good deeds through ten billion births, yogis obtain it and cross the ocean of existence.
3. Carefully press the perineum with the left heel, extend the right foot, and hold it firmly with the hands.
4. Lift up the haunches onto a seat, put the chin on the chest, close the nine [bodily] openings, and fill up the abdomen with air.
5. Put the mind at the crossroads[24] and commence breath control. Stop the movement of the Moon and Sun, and perform restraint of the breath.
6. This is the assimilation (*jāraṇā*) of impurity, the activation (*cāraṇa*) of *Bindu* and *Nāda*, the flushing of all the channels and the kindling of the fire.
7. Mastery of Body, Speech, and Mind by means of the Yoga of Body, Speech, and Mind is sure to arise through practice for the yogi who is on the path.
8. By means of this seal the yogi is sure to obtain everything, so he should zealously practise this Great Seal.
9. Because this self-born seal is the greatest of all seals, the best among the wise call it the Great Seal.
10. It checks death so is always beneficial. [Breath] control (*yantram*)[25] is mastered only by he who has fixed it in [his] mind.

[24] The microcosmic *catuḥpatha*, "crossroads," is a place where four channels meet. The location can vary; here, because of *Amṛtasiddhi* 12.12, which says that the lock at the perineum closes three channels, leaving open the central channel, it is likely to be at the bottom of the central channel, but in the *Tantrāloka* (15.94; cf. Jayaratha ad loc.) and *Mahānayaprakāśa* of Śitikaṇṭha (pp. 74 and 77) it is behind the top of the nose in the cavity above the palate.

[25] The referent of *yantra* here is unclear. In alchemical terminology *yantra* is a generic term for an alchemical device or crucible (Hellwig 2009: 18–23 describes and depicts several varieties of *yantra*. See also Sauthoff, Chapter 9 in this volume). It may also mean "body": verse 3c of the *Yogāvatāra* attributed to Nāgārjuna or Dignāga mentions the *dehayantra*; both the *Tantrāloka* and *Mokṣopāya* mention the *dehayantra* three times (*Tantrāloka* 28.220, 28.227, 28.230; *Mokṣopāya* 6.6.22, 6.30.65, 6.36.11); and Jayaratha *ad Tantrāloka* 28.230 understands *yantre* to mean "in the body" (*yantra iti dehe*).

11. The name spelled out by the first syllables of the words of its [name] is taught just for the amusement of those engaged in yoga.[26]

Thus ends the Inquiry into the Great Seal.

Chapter 12, the Inquiry into the [Great] Lock (*Mahābandha*)

1. Yogis are sure to master this Great Seal by means of [the technique] called the Great Lock, which keeps the breath in the body.
2. And the Great Lock should be known to be of two kinds, just as *Bindu* was said to be, for there is the Perineum Lock for the goddesses and the Throat Lock for the [great] god.[27]

[The Perineum Lock]

3. Inhale, hold the breath, perform the Great Seal, which removes danger, and quickly apply the [Perineum] Lock, which is kept secret from the gods and demons.
4. Simultaneously contract the anus and the perineum, make the Apāna breath move upward and join it with the *Samāna* breath.
5. The yogi should perform the [Perineum] Lock at the side of the opening of the Goddess of the Center. Having carefully blocked the three pathways, he should tightly restrain Breath.
6. Put Breath on *Prāṇa* and make *Prāṇa* face downward, and then after joining *Prāṇa* and *Apāna* propel [Breath] so that it moves upward.
7. This, the finest of unions in the body, illuminates the way of the Siddhas. Success, conviction, and well-being arise through only this and nothing else.
8. The channels, all of which usually flow downward, are reversed by this Great Lock.
9. A downwards flow makes the liquid elements leave the body, like rivers leaving their basins.

[26] This is obscure.

[27] At *Amṛtasiddhi* 2.5 the goddesses are said to reside at the lower end of the central channel and Śiva at its upper end.

10. Yogis should always know that this Lock functions in the body like a dam in the external [world] blocking a river.
11. This Lock is said [to work] on all the channels. By the grace of this Lock, the deities become manifest.
12. It is this Lock that, blocking three paths at the crossroads, opens the one path by which the Siddhas have attained happiness.
13. This downward path is said to bring about [pro]creation and death. The two [paths] that flow at [its] sides are agents of sin and virtue.

[The Throat Lock]

14. By putting all the elements into the pot of breath retention using a firm restraint and making *Udāna*[28] a [locking] bolt after joining *Prāṇa* and *Apāna*,
15. this [Throat] Lock stops upward movement in all the channels. By the grace of this [Throat Lock] alone is the Perineum Lock successful.
16. This is *Saṃpuṭa Yoga*. It is also taught as the Root Lock (*mūlabandha*). By means of this alone do good practitioners attain success in the three yogas.[29]

Thus ends the Inquiry into the Lock in the Attainment of Immortality.

Chapter 13, the Inquiry into the [Great] Piercing (*Mahāvedha*)

1. Know the [hand-gesture] seal to be, like the Lock, of two kinds: the Yoni Seal (*yonimudrā*) for the goddesses and the Penis Seal (*liṅgamudrā*) for the [great] god.
2. External creation arises from a woman and a man, but external [creation] from a woman and a man [ultimately] results in destruction of the bodily creation.
3. A woman of virtue and beauty is useless without a man; the Great Seal and Great Lock are useless without the [Great] Piercing.

[28] At *Amṛtasiddhi* 6.7 *udāna* is said to be in the region of the throat.
[29] i.e., the yogas of body, speech, and mind.

4. The yogi should take hold of the nexuses of Breath, perform the Lock as taught, and start performing the Piercing by means of Breath, which must be joined with Mind.
5. After greatly empowering *Prāṇa* [and] making it face the opening of the Goddess of the Center, he should use Breath to pierce the four stations in sequence.[30]
6. With the hands make the Penis [Seal] on the ground, [as unmovable] as the pole star, and point the two feet downward, making them just as stable.
7. Lift up the buttocks and, [staying] very steady, with the heels of the unbent feet [and] the pointed tip of the *vajra*[31] of the breath, flick the great Meru.
8. As a result Piercing takes place from the very bottom of the [microsmic] universe [upward]. The gods in the middle of Meru tremble because of Meru being made to move.
9. As a result of the Piercing of Meru, Brahmā and the other gods are sure to die. At first this Piercing happens quickly in the knot of Brahmā;
10. then, having broken through the knot of Brahmā, it breaks through the knot of Viṣṇu. Then, having broken through the knot of Viṣṇu, it breaks through the knot of Rudra.
11. Then, having broken the knot of Rudra and cut the creeper that is delusion, this Breath opens the very secret gateway of Brahmā.
12. The Great Seal, the Great Lock [and] the third, the Great Piercing: by means of these three elements, whose essences are secret, yoga is successful.
13. He who knows these three elements knows the three worlds. He who is lucky enough to practice them becomes omnipresent and [omni] potent.
14. It is because of this entitlement to secrets in the most secret Tantras that are secret [even] to the gods that humans alone have [the greatest level of] entitlement.
15. [By means of this practice] a man becomes anointed by all consecrations, endowed with all entitlements [and] situated on the path of the Siddhas. This cannot be otherwise.

[30] A system of four bodily stations (*catuṣpīṭha*) corresponding to four macrocosmic sacred sites Jālandhara, Uḍḍiyāna, Pūrṇagiri, and Kāmarūpa is taught in many tantric texts.

[31] A *vajra* ("thunderbolt") is a pointed weapon.

16. Pride is justified only for him who knows this seal: everything is understood by him thanks to the goddess that is the [central] channel.

Thus ends the Inquiry into Piercing in the Attainment of Immortality.

Chapter 14, The Inquiry into the Practice

1. He who has obtained this knowledge has obtained a wish-fulfiling jewel, has received a pot of goodness, a tree of plenty, a great treasure.
2. This secret triad of fundamental principles, the most important essence of the three worlds, does not bear fruit for men who are not devoted to practice.
3. In the same way that a treasure does not bear fruit for those who are not inclined to use it, so this [triad of] fundamental principle[s] will definitely not bear fruit for those who do not practice.
4. Having realized this, heroic men should practice regularly. Yoga results from regular practice, [and] as a result of yoga everything is accomplished.
5. After assuming the first seal and applying very firmly the two locks, make the three channels of the body sway.
6. And, using a firm application of the Male Seal,[32] flick the haunches after blocking the movement of the winds and inhaling and holding the breath.
7. In order to bring about an increase in all pleasures the yogi should perform the practice every three hours day and night without interruption.
8. By means of this practice Breath is trained. As a result of the training of Breath, the bodily Fire increases every day.
9. With Fire increasing, food is easily digested. Thorough digestion of food results in an increase of *Rasa*.
10. When *Rasa* is steadily increased the bodily constituents increase. It is as a result of an increase in the bodily constituents that the chief *Rasa* (semen) increases.

[32] i.e., the Penis Seal mentioned at 13.6.

11. When there is an abundance of the chief *Rasa* as a result of regular practice, the lord of yogis appears well nourished, with a firm body and great strength.
12. The Great Lock is sure to result from intensive practice. From intensive practice of the Great Lock there is assimilation (*jāraṇa*) of *Rasa*.
13. All the bodily humors (*doṣa*) and feces, urine, and [other] defilements dry up. [The chief] *Rasa* is made to move (*cāraṇa*) in the body through the channels at all times.
14. Through application of the practice the yogi can use the Yoga of Time to bring about stopping (*māraṇa*) [of Breath], unite the two [chief] *rasa*s, and [make Breath] enter the Goddess of the Center.
15. Having obtained a guru as helmsman and made a boat out of the body, men who cling to the oar of practice cross the ocean of existence.
16. Even if the gods themselves were to teach [this] wisdom again and again, [it is only] through practice that yoga manifests for yogis.
17. All the bad deeds done in thousands of lives are burnt up in this very life by means of the practice, like straws by fire.
18. The stated [rewards of] austerities, sacrifices, donations, and observances arise a hundred billion times over through the practice.
19. The various *Prakṛti*s, *Guṇa*s, *Doṣa*s,[33] and diseases in the body are destroyed by means of the practice, like clouds in the sky by the wind.

Thus ends the Inquiry into the Practice.

Bibliography

Primary Sources

Amaraugha. 2024. Edited by Jason Birch in The Amaraugha and Amaraughaprabodha of Gorakṣanātha: The Genesis of Haṭha and Rājayoga. Collection Indologie, 157. Haṭha Yoga Series 3. Pondicherry: Institut Français d'Indologie/École française d'Extrême-Orient.

Amṛtasiddhi and Amṛtasiddhimūla. 2022. Edited by James Mallinson and Péter-Dániel Szántó. Collection Indologie, 150. Haṭha Yoga Series 2. Pondicherry: Institut Français d'Indologie/ École française d'Extrême-Orient.

2001.1918–1938. *Dattātreyayogaśāstra*. 2024. Edited by James Mallinson. Collection Indologie, 162. Haṭha Yoga Series, 5. Pondicherry: Institut Français d'Indologie/École française d'Extrême-Orient.

[33] In the *Amṛtasiddhi* these are unwanted fluctuations in the body's natural order.

Haṭhapradīpikājyotsnā of Brahmānanda. 2002. Edited by Svāmī Maheśānand, Bāburām Śarmā, Jñānaśaṃkar Sahāy, and Ravindranāth Bodhe. Lonāvlā: Kaivalyadhām Kaivalyadhām Śrīmanmādhav Yogamandir Samiti.

Haṭhapradīpikā (10 chapters) of Svātmārāma with *Yogaprakāśikā*, a commentary by Bālakṛṣṇa. 2006. Edited by M. L. Gharote and P. Devnath. Lonavla: Lonavla Yoga Institute.

Haṭhatattvakaumudī. 2007. Edited by M. L. Gharote, Parimal Devnath, and Vijay Kant Jha. Lonavla: Lonavla Yoga Institute.

Hevajratantra. 1959. Edited by David L. Snellgrove. *The Hevajra Tantra: A Critical Study*, part 2: *Sanskrit and Tibetan Texts*. London: Oxford University Press.

Mahānayaprakāśa of Śitikaṇṭha in Old Kashmiri with a Sanskrit commentary. 1918. Edited by Mukunda Rāma Śāstrī. Kashmir Series of Texts and Studies 21. Bombay.

Matsyendrasaṃhitā. 2021. Edited by Csaba Kiss in Matsyendra's Compendium: *Matsyendrasaṃhitā* volume 1: *A Critical Edition and Annotated Translation of Chapters 1–13 and 55*. Collection Indologie 146. Hatha Yoga Series 1. Pondicherry: Institut Français d'Indologie/École française d'Extrême-Orient.

Mokṣopāya. 2018. Edited by Susanne Krause-Stinner and Peter Stephan. Das Sechste Buch: Nirvāṇaprakaraṇa, 1. Teil: Kapitel 1–119. Anonymus Casmiriensis Mokṣopāya: Historisch-kritische Gesamtausgabe, hrsg. unter der Leitung von Walter Slaje, Textedition, Teil 5. Wiesbaden: Harrassowitz.

Rasaratnasamuccaya of Vāgbhaṭa. 1890. Edited by Kṛṣṇarāvaśarmā Bāpaṭa. Ānandāśrama Sanskrit Series 19. Pune: Ānandāśramamudraṇālaye.

Rasārṇava. 1908–1910. Edited by Praphulla Chandra Ray. Bibliotheca Indica 175. Calcutta: Asiatic Society.

(*Śrī*)-*Jñāneśvarī* of Jñānadeva. 2001. Edited by G. S. Naṇadīkar. 5 vols. Mumbai: Prakāś Gopāl Naṇadīkar.

Tantrāloka of Abhinavagupta with commentary (°*viveka*) of Rājānaka Jayaratha. 1918–1938. Edited by Madhusūdan Kaul Śāstrī. Kashmir Series of Texts and Studies 23, 28, 30, 35, 29, 41, 47, 59, 52, 57 and 58. Bombay/Srinagar.

Vimalaprabhā = *Vimalaprabhāṭīkā of Kalki Śrī Puṇḍarīka on Śrī Laghukālacakratantrarāja by Śrī Mañjuśrīyaśas*. 1986–1994. Volume 1, edited by Jagannatha Upadhyaya. Bibliotheca Indo-Tibetica Series 11. Volumes 2–3, edited by Vrajavallabh Dwivedi and S. S. Bahulkar. Rare Buddhist Text Series 12–13. Sarnath, Varanasi: Central Institute of Higher Tibetan Studies.

Yogabīja. Forthcoming. Edited by Jason Birch. Collection Indologie, Haṭha Yoga Series. Pondicherry: Institut Français d'Indologie/École française d'Extrême-Orient.

Yogaprakāśikā, see *Haṭhapradīpikā* (ten chapters).

Yogaratnamālā, see *Hevajratantra*.

Secondary Sources

Digby, Simon. 2025. *Encounters with Jogīs in Indian Ṣ fī Hagiography*, ed. David Lunn. Delhi: Primus Books.

Fenner, Edward Todd. 1979. "Rasāyana Siddhi: Medicine and Alchemy in the Buddhist Tantras." PhD diss., University of Wisconsin–Madison.

Hellwig, Oliver. 2009. *Wörterbuch der mittelalterlichen indischen Alchemie*. Supplements to eJournal of Indian Medicine 2. Groningen: Barkhuis & University of Groningen, University Library.

Kvaerne, Per. 1977. *An Anthology of Buddhist Tantric Songs: A Study of the Caryāgīti*. Oslo: Universitetsforlaget.

Mallinson, James. 2014. "The Yogīs' Latest Trick." *Journal of the Royal Asiatic Society* 24 (1): 165–180.

Mallinson, James and Péter-Dániel Szántó. 2022. *The Amṛtasiddhi and Amṛtasiddhimūla The Earliest Texts of the Haṭhayoga Tradition*. Pondicherry: EFEO, Institut français de Pondichéry (IFP).

Pregadio, Fabrizio. 2019. *The Way of the Golden Elixir: An Introduction to Taoist Alchemy*. Mountain View, CA: Golden Elixir Press.

Sastri, K. A. Nilakanta. 1939. *Foreign Notices of South India: From Megasthenes to Ma Huan*. Madras: University of Madras.

Strnad, Jaroslav. 2013. *Morphology and Syntax of Old Hindī*. Amsterdam, Leiden: Brill.

White, David Gordon. 1996. *The Alchemical Body*. Chicago: University of Chicago Press.

Yang, Dolly. 2023. "Knowledge Transfer of Bodily Practices between China and India in the Mediaeval World." In *Yoga and the Traditional Physical Practices of India: Influence, Entanglement and Confrontation*, edited by Daniela Bevilacqua and Mark Singleton. *Journal of Yoga Studies* special issue 4: 413–440.

INDEX

For the benefit of digital users, indexed terms that span two pages (e.g., 52–53) may, on occasion, appear on only one of those pages.

Index written by Jolanta N. Komornicka

Note: An italicized *f*, *m*, or *t* after a page number indicates a figure, map, or table, respectively. Page numbers in **bold** refer to passages in translation. Glosses of Sanskrit terms are included at the main heading level when provided by the text. Where not, the English term has been kept separate from glossed versions.